GREAT WHALES

To the memory of
William Procter Bannister
1897–1963
and
Nancy Williams Bannister
1912–2005

AUSTRALIAN NATURAL HISTORY SERIES

GREAT WHALES

JOHN BANNISTER

CSIRO
PUBLISHING

© John Bannister 2008

All rights reserved. Except under the conditions described in the *Australian Copyright Act* 1968 and subsequent amendments, no part of this publication may be reproduced, stored in a retrieval system or transmitted in any form or by any means, electronic, mechanical, photocopying, recording, duplicating or otherwise, without the prior permission of the copyright owner. Contact CSIRO PUBLISHING for all permission requests.

National Library of Australia Cataloguing-in-Publication entry

> Bannister, J. L.
> Great whales/author, John Bannister.
>
> Collingwood, Vic. : CSIRO Publishing, 2008.
>
> 9780643093737 (pbk.)
>
> Australian natural history series
> Includes index.
> Bibliography.
>
> Whales – Australia – History.
> Whales – Australia – Identification.
> Marine mammals – Ecology.
> Whales – Conservation – Australia.
>
> 599.5

Published by
CSIRO PUBLISHING
150 Oxford Street (PO Box 1139)
Collingwood VIC 3066
Australia

Telephone: +61 3 9662 7666
Local call: 1300 788 000 (Australia only)
Fax: +61 3 9662 7555
Email: publishing.sales@csiro.au
Web site: www.publish.csiro.au

Front cover
Humpback whale breaching off Queensland's coast. Photo by Joshua Smith and Michael Noad, Cetacean Ecology and Acoustics Laboratory, University of Queensland.

Back cover
A dwarf minke whale in the Great Barrier Reef. Photo by Alastair Birtles
© James Cook University Minke Whale Project.

Set in 10.5/14 Adobe Palatino, Optima and Stone Sans
Cover and text design by James Kelly
Typeset by Desktop Concepts Pty Ltd, Melbourne
Printed in Australia by BPA Print Group

CONTENTS

Acknowledgments		vii
Preface		ix
1	Introduction	1
2	Fascinating creatures	5
3	Whales and people	13
4	Biology	27
5	Species accounts	49
	Blue whale	50
	Fin whale	56
	Sei whale	59
	Bryde's whale	62
	Minke whale	73
	Humpback whale	77
	Southern right whale	89
	Sperm whale	98
6	Conservation and regulation	105
7	The future	115
Glossary		123
Further reading		135
Index		139

A breaching southern right whale off the Western Australian south coast. Photograph by Ray Smith © DEWHA

ACKNOWLEDGMENTS

Help, advice and encouragement have come from many people over the years since my first encounter with 'the whales'. For initial encouragement and guidance I must thank Neil Mackintosh, Sidney Brown, Malcolm Clarke, Geoffrey Kesteven, Graham Chittleborough, Bill Dawbin and Geoff Kirkwood. For friendship, support and tolerance over a range of years I am particularly grateful to Peter Best, Bob Brownell, Steve Burnell, Martin Cawthorn, Greg Donovan, Ray Gambell, Durant Hembree, Sharon Hedley, Cath Kemper, Vicky Rowntree and Bob Warneke. In early years at the Western Australian Museum David Ride encouraged me to maintain my interest in whales, including, from 1970, annual participation in the IWC Scientific Committee.

A public lecture I give on Western Australia's Great Whales (southern right, humpback, blue and sperm) has 74 names listed on the acknowledgment slide. While not attempting to repeat all the names here, I must mention those who have over the years undertaken the southern right, humpback and blue whale surveys off Western Australia, either as pilots, observers or photographers, especially John Bell, Chris Burton, Kitch Godfrey, Mich Jenner, Ray Smith, Alan Murdoch, Andrew Halsall, Jenny Schmidt, Gail Neylan and Julie Biser.

In compiling the information for this book I have received help, directly or indirectly, from numerous people, including Cherry Allison, Alastair Birtles, Peter Best, Steve Burnell, Chris Burton, Doug Coughran, Greg Donovan, Paul Ensor, Stuart Frank, Nick Gales, Jason Gedamke, Pete Gill, Tatjana Good, Nicky Grandy, Curt and Micheline Jenner, Cath Kemper, Rob McCauley, Robyn McCulloch, Mike Noad, Bill Perrin, Milena Rafic, Randy Reeves, Vicky Rowntree, Jon Seger, Tim Smith, Steve Van Dyck and Bob Warneke. Rob McCauley kindly read the section on sound production in Chapter 4, and the paragraphs on acoustics in the various species accounts in Chapter 5; he provided Figure 4.5.

I am specially grateful to Martin Thompson for the species illustrations and to Chris Burton, Doug Coughran, Paul Ensor, Curt Jenner and Paula Olson for providing photographs. Special thanks go to Nick Alexander, particularly for his forbearance in unravelling my overlong sentences.

Permission to reproduce Figure 3.2 has been granted by the New Bedford Whaling Museum, for Figures 5.4 and 5.13 by the Editor, *Marine and Freshwater Research*, for Figure 5.6 by the Editor, *Memoirs of the*

Queensland Museum, and for Figure 5.11 by the Head of Science, International Whaling Commission.

Cherry Allison of the International Whaling Commission provided the catch details for Tables 5.1 and 5.2.

Steve Van Dyck of the Queensland Museum and the relevant authors have allowed me to consult pre-publication versions of their articles in the third edition of *The Mammals of Australia*. Randy Reeves and his colleagues have also allowed me to use, as background, documents originally prepared by Justin Cooke and discussed at an IUCN workshop in January 2007 to assess the global status of cetacean species; since the decisions of that workshop have yet to be ratified the existing assessments have been used here.

I remain specially indebted to the Trustees, Chief Executive Officer and staff of the Western Australian Museum who have provided administrative and other support at the Museum over many years. Particular debts are owed to Di Jones, Executive Director, Collections and Content Development, Anne Nevin and Grefin Harsa, her past and current personal assistants, and to Margaret Triffitt, Museum Librarian and Wendy Crawford for many kindnesses.

Lastly I want to thank my family, particularly Kate, John and Caroline, for their support, and Yass, for being there over the past seven years.

PREFACE

> *'Would you like to go to the Antarctic?'*
> *'Yes ... please ... er ... when? And for how long?'*

It all began with that snatch of a conversation, in the morning coffee queue. The questioner was the Zoology Professor, Sir Alister Hardy, and we were where all business was transacted in his Department every morning from about 10.30 during term time. Professors, students, technicians, lecturers, all queued up, to gossip, exchange erudite quips, arrange tutorials, enquire about the state of that weekly essay, commiserate over the Test score (Australia was always winning) ...

And that morning 48 years ago it was my turn. In his inimitable way, head slightly tilted like a blackbird listening for worms, he was approaching, arm outstretched, and eyes glinting behind those round spectacles.

'Well, now. Write to Dr Neil Mackintosh at the National Institute of Oceanography. He's looking for people like you. My secretary will give you his address. And don't forget, it's Dr N. A. Mackintosh ... CBE ...'

And so my life was arranged for me for the foreseeable future. It involved six months on the island of South Georgia, in the South Atlantic, as a Junior Whale Fishery Inspector (no less), employed by the Falklands Islands Government at the Grytviken whaling station of the Compania Argentina de Pesca – Argentine registered, managed by Norwegians, with a mixed labour force of Norwegians, Uruguayans, British and others.

The result was an encouragingly large collection of whale bits and pieces, mostly in formalin in old salt beef casks, and an offer to work at NIO (as the UK National Institute of Oceanography was fondly called) thereafter. And three years later, in April 1964, the wheel turned again, and I found myself, four-months married, in Perth on a three-year contract with CSIRO to work on sperm whales then being taken at Albany, on Western Australia's south coast. And one way or another I've been in Australia, involved with whales, more or less, ever since – in fact, originally composing these words on the day after my seventieth birthday, I wondered how it all happened, and reminded myself how much my life owes to the whales.

An interacting group of 'unaccompanied' southern right whales, Western Australian south coast. Photograph by Andrew Halsall © DEWHA

1
INTRODUCTION

And God created great whales.
Genesis, chapter 1, verse 21, first quoted in Herman Melville's
'Extracts' at the beginning of Moby-Dick, *1851.*

The word 'whale' is usually taken to mean the larger members of the mammalian order Cetacea – the whales, dolphins and porpoises. That classification into the three groups is loosely based on size. The term 'whale' generally refers to cetaceans over about 10 metres long (although two of the smallest, the dwarf minke and pygmy right whales, grow only to about seven metres, while others, such as pilot whales and some beaked whales, can be even smaller); 'dolphins' are between about two and five metres long; and 'porpoises' below two metres. But there is no real biological basis for these different names.

Zoologically speaking the Cetacea are divided into two suborders, based on their feeding apparatus: the baleen whales, the Mysticeti, and the toothed whales, the Odontoceti. The baleen (or 'moustached') whales use baleen plates in the mouth to filter their food from sea water, while the toothed whales all possess teeth, even though in some species the teeth are extraordinarily modified and reduced in number. And while all the mysticetes are 'whales', only relatively few odontocetes are whales, such as the sperm and killer whales, the weirdly modified beaked whales, the solely arctic narwhal and white whale, the pilot whales, and one or two others. The rest of the odontocetes comprise the dolphins and porpoises.

Of the two suborders, the odontocetes are much more numerous, with eight families, 34 genera and some 73 species. There are only four families of mysticetes, comprising six genera and 14 species. By world standards, Australia's cetacean fauna is moderately rich. It has representatives of eight of the world's 12 families, 27 of the 40 genera, and 44 of the 87 or so species currently recognised.

Only one of our cetaceans is endemic – the recently described snubfin dolphin, *Orcaella heinsohni*, from Townsville, Queensland. However, Australia is the type locality for that and three other species – the pygmy right whale, *Caperea marginata*, first described from three baleen plates collected at the Swan River Colony, Western Australia; the southern bottlenose whale, *Hyperoodon planifrons*, collected as a beach-worn skull from the Dampier Archipelago, Western Australia; and the tropical bottlenose whale, *Indopacetus pacificus*, described from a skull collected from the beach at Mackay, Queensland. Until recently the tropical bottlenose whale was quite unknown as a living animal.

Four of the world's cetacean families are not found here. The absent ones are the northern hemisphere gray whales (family Eschrichtiidae, now found solely in the North Pacific); the Arctic narwhal and beluga (family Monodontidae); and two families of the largely freshwater river dolphins found in South America and eastern Asia. None of the porpoises are found regularly near the Australian continent. The single 'Australian' representative, the spectacled porpoise, *Phocoena dioptrica*, is an Antarctic animal that occasionally strays north into our waters. The remaining seven families are well-represented here. This is not surprising given Australia's wide range of coastal habitats – southern cool temperate to northern tropical – and its position in the path of migration routes of such regular ocean travellers as the humpback and southern right whales.

Not unexpectedly, Australia shares many cetacean faunal elements with the two other southern continents and New Zealand. But it lacks one group: the coastal dolphins of the genus *Cephalorhynchus* (the South American Commerson's and Chilean dolphins; Haviside's dolphin of south-west Africa; and Hector's dolphin of New Zealand).

Only three cetaceans have been sufficiently numerous close to the Australian coast to be commercially important here: the southern right whale, the humpback whale and the sperm whale. This contrasts with the

Table 1.1: Great whale species dealt with in this book

Order	Suborder	Family	Species/Subspecies
Cetacea (whales, dolphins and porpoises)	Mysticeti (baleen whales)	Balaenidae (right whales)	*Eubalaena australis* (southern right whale)
		Balaenopteridae (rorquals)	*Balaenoptera musculus intermedia* (Antarctic blue whale)
			Balaenoptera musculus brevicauda (pygmy blue whale)
			Balaenoptera physalus (fin whale)
			Balaenoptera borealis (sei whale)
			Balaenoptera edeni (Bryde's whale)[1]
			Balaenoptera bonaerensis (Antarctic minke whale)[2]
			Megaptera novaeangliae (humpback whale)
	Odontoceti (toothed whales)	Physeteridae (sperm whales)	*Physeter macrocephalus* (sperm whale)

1 See 'Chapter 5, Bryde's whale' for current status of *Balaenoptera omurai*
2 See 'Chapter 5, Minke whale' for current status of the 'dwarf' minke

situation off southern Africa, where eight species of whale (blue, fin, sei, Bryde's, minke, humpback, southern right and sperm) have all been subjected to coastal whaling. Interestingly enough all these species, except for the tropical/temperate Bryde's whale, have also been subjected to whaling in the Antarctic, well south of Australia.

That brings us to the 'great whales', the subject of this book (see Table 1.1). Here too we are not talking zoologically, but the reference is to the larger whales, almost all once commercially important, and several still highly significant in the context of the Australian fauna. Strictly speaking the 'great whales' comprise the six baleen whales (blue, fin, sei, Bryde's, humpback, southern right) and the one toothed whale, the sperm whale. But I have added another – the minke – not quite a 'great whale' in the traditional sense, but important enough in the Australian context to be included here. Not only does it have two forms – the dwarf minke, subject of a popular commercial 'swim' program off northern Queensland, and the Antarctic minke, common in summer in the Antarctic – but the latter is frequently in the news because it has been, and continues to be the main (and controversial) target of 'scientific permit whaling' in the Antarctic south of Australia. So minkes, of both forms, are included here.

Some explanation is necessary here about the terms 'baleen whale', 'balaenopterid', and 'rorqual'. The 'baleen whales' are all those species that make up the suborder Mysticeti. The 'balaenopterids' are all the members of the family Balaenopteridae, including the six members of the genus *Balaenoptera* (blue, fin, sei, Bryde's and the dwarf and Antarctic minke whales) and the single member of the genus *Megaptera* (the humpback). Those seven species (assuming the dwarf minke to be separate) are the 'rorquals'. Professor Lars Walløe of the University of Norway has shown that the word rorqual comes from the Norse *rørkval*, from an earlier word *reydr* meaning 'tubes or grooves', referring to the ventral grooves of Balaenopterids. The word *rørkval* became *rorqual* first in French, then English. Strictly speaking 'rorqual' probably should not include *Megaptera*, the humpback, but for present purposes, it will be taken to mean any of the species in the family Balaenopteridae, which includes the humpback. Right whales, as members of the separate family Balaenidae, are not rorquals, and do not have ventral grooves.

And why just Australia's 'great whales'? First, because they make a sufficiently coherent and convenient grouping for treatment together, and second, because many of these elusive and graceful creatures have been, or are, specially important to Australians.

For the early settlers, whales were a major feature of the landscape – right whales in the Derwent River, Hobart, for example. In the early days, right whales and sperm whales were taken by visiting Yankees, British or French, as well as by the colonials. In more recent times, humpbacks were the target of the whaling industry based at coastal stations in Western Australia, Queensland and New South Wales, just after World War II, while illegal Soviet operations in the 1960s took humpbacks, sperm and pygmy blues. These days, the highly popular whalewatching industry is based around humpbacks, right whales, blues, and dwarf minkes.

Most particularly, whales are worth writing about not just because the blue whale is the largest animal that has ever lived, or the sperm whale is the biggest and possibly most bizarre of the toothed whales, but because they include the humpback and southern right whales, both of which have for some years been coming back from the brink. As the largest and most iconic visitors to Australian shores the great whales deserve our admiration, respect, and continued care. It is my hope that the information gathered in this book will help achieve those aspirations.

2
FASCINATING CREATURES

So is this great and wide sea, wherein are things creeping innumerable, both small and great beasts. There go the ships: there is that leviathan, whom thou hast made to play therein.
Psalms 104: 25, 26; quoted by L Harrison Matthews in
The Whale, *1968.*

Since the earliest days of natural history … the … whale has been subjected to constant misrepresentation … [as a result] of that heated imagination which leads some enthusiasts to see nothing in nature but miracles and monsters.
Baron Georges Cuvier, quoted by Thomas Beale in The Natural
History of the Sperm Whale, *1839.*

Few people nowadays would have much difficulty telling you that a whale, though it lives in the sea, is not a fish but a mammal, like its smaller relatives the dolphins and porpoises. Nor that Australia once used to kill whales commercially – but of course that was a while ago. Nor that now you can see whales in spectacular David Attenborough documentaries, or on the Discovery Channel, and that you can go whalewatching and experience them as real live animals, out there at sea, amazing, friendly, inquisitive creatures … literally in their element.

It wasn't always quite like that. Aristotle knew they were mammals – they had lungs, dolphins snored (through a blowhole), whales and dolphins gave birth to live young and fed their young on milk. But despite Aristotle's wisdom, most ancients thought whales were mythical creatures, literally monsters of the deep. Jonah was swallowed by 'a great fish' – in those days equivalent to a whale. Leviathan (though originally probably a crocodile) was any absolutely enormous creature, so whales were leviathans.

In mediaeval times, the Scandinavians and Icelanders wrote a lot about whales – a thirteenth century volume even says there's not much more than whales worth writing about (at any rate in Norway). Some whales were kind and helpful – minke whales used to help fishermen by driving fish shoals inshore, and fin whales were good to eat. A mediaeval Norwegian document (translated by Prof. Lars Walløe of the University of Oslo) states that the sperm of a whale 'if you be sure that it came from this sort and no other … will be found a most effective remedy for eye troubles, leprosy, ague, headache and … every other ill that afflicts mankind'. Other whales, though, were monsters, and dangerous. And because artists had often never actually seen a whale, their illustrations tried to make them look as fearful as possible.

Occasionally a whale would strand on the shore, and cause great interest. It has been said that old engravings often exaggerate the size of such creatures by drawing the people around it as dwarfs, so the animal seems even more monstrous than it really is. Even now, as Leonard Harrison Matthews says in his book *The Whale*, old beliefs about whales die hard. Terms such as whalebone (not bone at all), ambergris (certainly not amber) and spermaceti (nothing whatever to do with reproduction), are still in common use.

Whales are still seen as mysterious, fascinating creatures. Despite modern technology, they inhabit a world still largely unexplored and unknown. They can only be seen, or rather glimpsed, when they are near the sea surface, either from boats, or perhaps from shore, or underwater by divers or from submersibles. They include, in the blue whale, the largest animal ever to have lived on the planet. That huge animal can grow to 30 metres in length – that's equivalent to the height of a six storey building. A blue whale can weigh more than 130 tonnes – more than the weight of 20 African elephants. Even the largest dinosaurs may have weighed not quite 100 tonnes. And yet, despite its vast size, the blue whale feeds on individual creatures only a few centimetres long, although it does take in a huge quantity of them at a time, perhaps more than four tonnes a day.

Superbly and wonderfully specialised, whales – as well as the dolphins and porpoises – spend their entire lives in water. Among the marine mammals, only the dugong is so independent of land. Seals, sea lions, walruses, sea otters (usually) all need to come on to land at some time in their lives, mainly to reproduce.

That life spent entirely in water has so shaped their lives and so altered their bodies that cetaceans seem not to be mammals at all. They have had to overcome the physical resistance of water – hence streamlining, and loss of hair, and smoothness of skin. Water also conducts heat, so they have to insulate themselves. Water also conducts sound very well, and they have very special adaptations to take advantage of that.

As good mammals, whales, dolphins and porpoises all need to take in air at the surface, and to do that they have developed the ability to empty the lungs very rapidly through their blowhole. A large whale can exhale rapidly, taking only one to two seconds, and this is followed by an equally rapid intake of air. Once often known as the 'spout' and now more commonly (and accurately) as the 'blow', the blow may be the first, and sometimes only, visible evidence of a whale's presence at sea. Blue whales have a tall and columnar blow; humpbacks have a 'bushy' blow; the blow of sperm whales is short and directed diagonally forwards while that of right whales is V-shaped. The blow's visibility may to some extent be due to a small amount of water carried up from the blowhole, but is more generally believed to be due to the cooling effect of air released under great pressure. Even in the tropics, the cooling effect causes condensation of water vapour and the 'spouting' appearance.

Among the cetaceans, two rather different life styles have evolved within the two distinct suborders: the mysticetes – whalebone or baleen whales; and the odontocetes – toothed whales, dolphins and porpoises. The relatively few baleen whale species contrast markedly with the much more numerous odontocetes. Most of the mysticetes are very much larger in size. Only one odontocete, the sperm whale, at a maximum length of around 18 metres (and then only in the largest males) even approaches the larger baleen whales. The next biggest odontocetes – killer whales and some beaked whales – are full grown at around nine metres and only exceed the size of the very smallest baleen whales.

In other ways, too, the baleen and toothed whales are very different from each other. For one thing, baleen whales are grazers, taking in vast quantities of their food at a gulp, while toothed whales actively chase their individual prey. And each group has a different life style. Baleen whales, even though they are social animals, have a rather simple social life, going around in twos or threes, usually gathering in large groups only to feed. Toothed whales lead much more socially active lives, in highly organised family groups that may last a lifetime, with strong social bonds, often in matriarchies. Sperm whales have been compared to elephants in their long lives, large size, matriarchal social structure and breeding systems. This leads to the question of how 'intelligent' such animals may be.

The question of intelligence is often raised in conversations about cetaceans but there is no simple answer. Professor Berndt Wursig's entry on the subject in the *Encyclopedia of Marine Mammals* occupies over eight pages of close double-column type. But while each group is marvellously adapted to its own lifestyle – and in Professor Wursig's word 'smart' – baleen whales seem less adaptable, perhaps more like cattle, than most toothed cetaceans, which one can liken perhaps to dogs in their ability (at least in dolphins) to learn tricks, and which demonstrate complex social structure and behaviour. It would be fascinating, but obviously impractical, to study a captive sperm whale, but this toothed whale *par excellence*, despite its size, comes out considerably higher in one objective measure – percentage brain size relative to body weight – than, for example, the fin whale. For the sperm whale the figure is around 0.023 – three times more than the fin whale's 0.008. But, as Professor Wursig concludes, admittedly for dolphins (but for which one might perhaps just as well read sperm whales):

> [They] are not those 'super intelligent' beings as claimed by some ... and ... are indeed 'intelligent' for those things that they need to solve and interact with in their natural world ... [a] world ... very different from ours.

Another often-posed question is 'why do whales strand?' Strandings generally involve one of two kinds of animal – singletons, or 'mass' strandings. Some animals, usually single, are either dead or dying before they reach the shore. It has been said that sick individuals, in an attempt to keep breathing, may have some instinctive urge to beach themselves. Even when returned to the water they come ashore again. Such animals are often brought dead or dying on to the beach on a high tide, or by an onshore current, or with a strong onshore wind. Mass strandings, however, usually involve animals very much alive when they arrive on the beach, and the

Figure 2.1 Stranded sperm whales at Cheynes Beach, south coast of Western Australia, October 1969. The enormity of any rescue in such circumstances is obvious. Photo © Penny Bannister

reasons for their stranding are often unclear. Post-mortems have found little if any pathological evidence for the occurrence, and a number of ingenious theories have been put forward in explanation, including the possibility that cetaceans may use the Earth's magnetic field for navigation and that magnetic disturbances or anomalies may have disorientated them. The possibility that strandings of beaked whales, in particular, may have been caused by seismic (including military) activities has caused recent concern. The most likely explanation for mass strandings, however, seems to lie in a combination of factors, including the following:

- the species involved are usually those open ocean animals (including sperm whales) that rarely otherwise approach the coast
- such animals are usually highly social, and may naturally group close together in case of danger, even going to an injured companion's aid
- on the few occasions when mass strandings have been observed in progress, a 'lead' animal may have come ashore first
- stranded animals are often grouped very close together on shore and, when returned alive to the water near animals still on shore, 'rescued' animals tend to return immediately to the stranding site.

Thus such strandings occur where an otherwise deep-water species finds itself close to the coast in what are presumably unfamiliar circumstances. One or more may get into difficulties and become stranded

Figure 2.2 A group of volunteers attempt to refloat a very young humpback whale off the coast of Fraser Island, Queensland. Photo © Oryx Publishing

in the shallows. The rest of the group instinctively crowd together, those in distress presumably communicating their distress to the others. As the tide goes out, those on the beach overheat very rapidly, and die from that, or lung collapse, or stress, or a combination of such factors. Any animals that do get back into deeper water, but within sound of those still alive in the shallows, will return to join them. Successful rescues, often involving large numbers of volunteers, helped by wet suits and surf skis, not to mention considerable coordination, depend upon removing the animals well away from the stranding site. Rescues are possible for those smaller species (but regrettably not for larger ones such as sperm whales, see Chapter 5, Sperm whale) where it is practical to remove them from the beach. Well-developed protocols are available at a State level to handle such events.

There is now some evidence that mass strandings may occur in cycles. In Tasmania, such events are particularly frequent: Professor Mark Hindell and his colleagues at the University of Tasmania have demonstrated an approximately 10-year frequency in mass strandings which is linked to periodicity in weather patterns bringing food further north than usual. And off the south coast of Western Australia an uncanny pattern of such strandings seems to have been developing over the past 40 years: in the 1960s, most were recorded along the stretch of coast well east of Albany; in the 1970s there was at least one notable stranding (of short-finned pilot whales) near Denmark, west of Albany; and in the 1980s there were several well-publicised events (two involving highly successful rescues) on the continent's south-west corner, near Augusta. And now in the early years of the twenty-first century there have been several mass strandings once again east of Albany. There may be nothing in it, but it raises intriguing possibilities.

To round off this excursion into cetacean marvels, two products, ambergris and scrimshaw, often catch the public imagination, although neither has been of great importance on the scale of whale oil, whale meat products or even whalebone.

Ambergris is formed as a concretion in the intestine of the sperm whale. The name comes from the French, *ambre gris*, meaning grey amber, as distinct from the true amber, *ambre jaune*, or yellow amber, which comes from pine tree resin. When fresh it is blackish-brown, sticky, and evil smelling, but when exposed to air it gains a sweet, musky and not unpleasant odour. It is then solid and friable, described by Dr Dale Rice of the National Marine Mammal Laboratory, Seattle, USA, as having the consistency of 'nearly dry clay', pale yellow to light grey inside and dark brown and varnished-looking on the outside. When ambergris is broken apart, individual lumps tend to fracture along concentric lines. Almost invariably squid beaks are found embedded in the lumps. During the days of sperm whaling, a careful lookout was kept for it – it was recorded in about one to five per cent of animals examined. Although most pieces are fairly small, weighing between 0.1 and 10 kg, some very large lumps have been found. The largest, recorded by Dr Robert Clarke of the then UK National Institute of Oceanography from a male in the Southern Ocean in December 1953, weighed 420 kg and measured some 1.7 by 0.8 metres. Another, from a sperm whale killed off the Australian coast before World War I, is said to have been so large that when sold the proceeds saved the company from bankruptcy. However, ambergris generally occurs in much smaller pieces. It was once said to be more than worth its weight in gold, and was prized in the East for its reputed aphrodisiac properties, and even used in cooking and medicine, although its main feature and value lay in its ability to 'fix' the perfume of more desirable items, such as exotic roses. Nowadays it has been largely replaced by synthetics. How it forms is unknown, though it seems to occur in response to some irritant, hence the presence of squid beaks – which don't normally pass into the sperm whale intestine. Ambergris floats in water and may occasionally be found on the beach. It is subject to the provisions of the *Environment Protection and Biodiversity Conservation Act 1999* (see Chapter 6), which regulates the export and import of whale products, and any finds should be reported to the nearest State or Territory environment office. An example of beach-found ambergris is pictured in the colour section, page 63.

Scrimshaw is more properly an artifact than a whale product *per se*. It is the handiwork created by whalers, most commonly from the teeth or bones of sperm whales or the baleen of other whales. As described by Dr Stuart Frank of the New Bedford Whaling Museum, Massachusetts, the 'quintessential manifestation of scrimshaw' was 'pictorial engraving on sperm whale teeth', often naive, but sometimes remarkably sophisticated, and highlighted with lampblack or sealing wax. That is how most people know it today but it also took other forms, where not only teeth but sperm whale bone or right whale baleen might be fashioned into things as diverse as walking sticks, 'swifts' for winding wool, sewing implements, pie crust crimpers or decorative plaques. Typically the product of whalers while 'skrimshanking' (a term not necessarily referring to activities during idle periods on board but also to making articles for ship's use, such as tool handles, 'fids', belaying pins), the practice flourished during 'traditional' sperm whaling from the 1830s, and has persisted in various forms to the present day. Genuine articles now fetch high prices, and fakes are common, not only as resin or plastic replicas, but as genuine teeth or bone doctored to appear old. A small collection of representative items is shown in the colour section, page 63.

Modern scrimshaw from the teeth of sperm whales caught after Word War II is still available in Albany or Perth. These items may be bought and sold within Australia, although all such 'whale ivory' is subject to the import and export provisions of the *Environment Protection and Biodiversity Conservation Act 1999*, and internationally under regulations of the Convention on International Trade in Endangered Species (CITES) which covers trade in endangered species and their products.

3
WHALES AND PEOPLE

The mighty whales which swim in a sea of water, and have a sea of oil swimming in them.
  Fuller's Profane and Holy State, quoted in Moby-Dick, *1851.*

[The right whale] does no harm to ships: it has no teeth, and is a fat fish and well edible.
  From the 13th century Norwegian Speculum Regale, quoted in L. Harrison Matthews, The Whale, *1968.*

The sovereign'st thing on earth was parmaceti for an inward bruise.
  Shakespeare, quoted by Frank T Bullen,
  The Cruise of the Cachalot, *1898.*

Whaling

Whaling is said to have begun in or about the 11th century with the Basques in the Bay of Biscay, from 'their stronghold in the crook of the elbow of the Iberian peninsula' as described by the author Richard Ellis. Although

Norsemen were hunting whales off the coast of Scandinavia before that, the Basques seem to have been the first to do so in any kind of organised way. From the Biscayan coast, they ventured out across the Atlantic towards Labrador and Newfoundland, perhaps via Iceland. On both sides of the Atlantic the Basques were taking right whales, but they may also have been taking the Atlantic gray whale – it became extinct by around 1700. The Basques then transferred their activities north towards Spitzbergen (now Svalbard) in the Arctic, in pursuit of the 'Greenland' whale – nowadays regarded as a relict population of the Arctic bowhead. They were soon out-competed by the Dutch, who then, and later with the British, pursued the 'Greenland' whale further west so that it was virtually extinct before the end of the 19th century. At the same time, from around the mid-1600s until the early 1800s, whales were being taken by the Japanese, in coastal net fisheries. Unlike the Basques, Dutch and British, the Japanese used the whole whale, not just the blubber or baleen, anticipating by some hundreds of years the 'modern', all-embracing industry, still today being pursued by them (see Chapter 6).

The end of 'traditional' Japanese coastal whaling has been attributed to the arrival of Yankee whalers searching for sperm whales nearby in the North Pacific. The flowering of the Yankee, worldwide, 'open boat' industry in the first half of the 19th century, dates from small beginnings off New England early in the eighteenth century, with the first sperm whales caught from about 1720, between Cape Cod and Bermuda. That was followed by a gradual spread throughout the Atlantic until the sperm whalers moved into the Pacific, around Cape Horn from 1786, and into the North Pacific by 1818. Whaling had begun in the Indian Ocean, presumably via the Cape of Good Hope, by at least 1791. The American sperm whale fishery climaxed in 1837, based on the 'exceedingly profitable' Pacific grounds, but by the late 1880s, whaling had ceased in many formerly very successful areas, and others were by then seldom visited. Along with the Americans, British, French, German and some other nationalities also took part; as well as sperm whales, right whales, and to a lesser extent humpbacks, were the target.

Of particular interest in the Australian context are early 19th century developments in the south-west Pacific and the Indian Ocean. On the western side of the Pacific a New Zealand sperm whale fishery, dominated by British whalers, began in 1802. Whaling then expanded, peaking in the 1830s, when American whalers took over. However, their presence soon declined very rapidly – from a total of 108 visits in 1839 to only 35 in 1841.

A colourful description by the popular author Ivan Sanderson illustrates the advance of 'spermers' through the Pacific from Cape Horn along the coast of South America, west along the equator ('on-the-line'), then branching south towards New Zealand as well as north towards Japan. 'In a matter of two decades the Pacific, from being a vast *aqua incognita*, became a [veritable] whalers' pond.'

In the Indian Ocean sperm whaling again peaked around 1840. At that time there were five principal Indian Ocean sperm whaling grounds: two off the east African coast around Madagascar, one off the Arabian coast, another off the west coast of Java, and one off the north-west and south coasts of 'New Holland' – as Australia was then known.

A review of 146 American logbooks available on microfilm for sperm whaling on the 'Coast of New Holland Ground' for the period 1835–1875 showed that by far the majority (75 per cent) covered the period to 1859, confirming the peak of activity there to that time. The rapid growth of the 'New Holland' sperm whale fishery was aided by the proximity of Australian ports to those highly productive cruising grounds, enabling ships fitting out in those ports to perform three voyages while others could only perform two.

Right whaling was pursued on the Australian south-west coast in winter. There was then a period of summer sperm whaling off the north-west coast, towards the East Indies (Indonesia), followed by a return for more right whaling in winter. After more early summer right whaling towards the west (and possibly some sperm whaling, too), the whalers returned home to American ports the same way as they had come – via the Cape of Good Hope and up through the Atlantic. But the pattern changed after 1843. Vessels then passed through the 'Coast of New Holland' whaling ground *en route* for Van Diemen's Land (Tasmania). During January and February they would whale there (presumably taking sperm whales), before sailing north-east through the Pacific to the then prolific 'North West' right whaling grounds of the USA. From there they would return home via Cape Horn.

The peak of whaling activity in the Indian Ocean around 1840 was followed by a gradual decline towards the century's end. A contemporary account states: 'The number of American vessels visiting the … ocean has been gradually diminishing for several years, and in 1880 not a single vessel from the United States went there for sperm oil …' although sperm whales could still be encountered. There is reference to Tasmanian vessels continuing to sperm whale 'with good success' in some years, but probably by then only in the south-west Pacific.

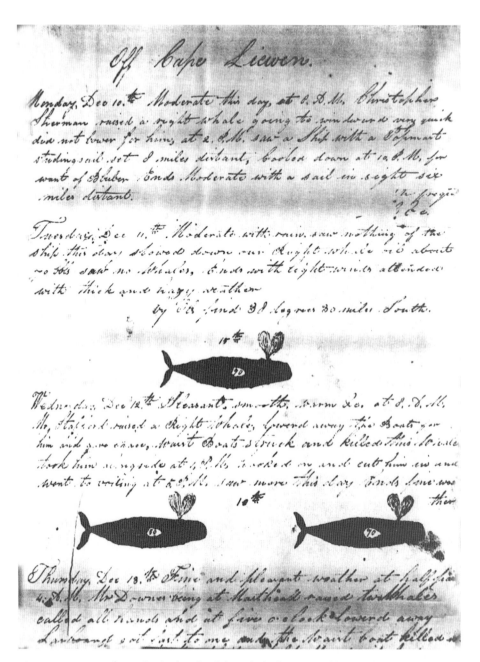

Figure 3.1 A page from the logbook of the whaleship *Emerald* records three right whales caught off Cape Leeuwin, Western Australia, in December 1838. The small ink stamps show the number of barrels of oil obtained from each whale.

Source: from Pacific Manuscripts Bureau microfilm 209, held in the State Library of Western Australia

Whales and people

Figure 3.2 Two pages from the journal of William Wells Eldridge on the barque *Canton*, cruising off the Western Australian south coast between Bald Head, Albany and Cape Chatham, January 1876. Included is a sighting of sperm whales.

Source: Reproduced courtesy of the New Bedford Whaling Museum

'Open boat' whaling was, as its name indicates, undertaken in lightly constructed boats, no more than about nine metres in length, and (in the American fishery) crewed by six men. Up to four of these small, undecked boats would be carried on the parent whaleship. The whaleship itself was a relatively slow but well-found vessel, described as 'a sturdy, bluff-bowed, flat-bottomed sailer'. Typically between 30 and 45 metres in length, they would cruise the whaling 'grounds' (localities recognised as the haunts of sperm or right whales), with lookouts in their masthead hoops, scanning the ocean for a sign of the quarry. When a sighting was made (the lookout shouting 'Blows' or 'She blows') the whaleboats would be lowered for the chase and rowed or sailed up to the whale. The whale would then be 'fastened' by a harpoon (known as an 'iron') attached to a long whale line. The whale might then sound, or swim rapidly at the surface. In due course

Figure 3.3 An 'open boat' whaling scene in the early 19th century illustrated the rigours of the chase and the death of a sperm whale. In the background are shown the removal of the blubber 'blanket piece' and the boiling out of the oil on the whaleship's deck. Engraved from a painting by the French artist Garneray

the whale, now tired and back at the surface, could be approached once more, this time for the kill, when a long, razor-sharp lance ('killing iron') would be thrust into its body – preferably into the lungs or heart (its 'life') – and 'churned' until its final 'flurry', after which the carcass had to be towed back to the whaleship for processing.

Processing consisted essentially of tearing off the blubber from the carcass moored alongside the whaleship. It was done in a spiral, rather like unwinding a bandage. The blubber, in 'blanket pieces', would then be cut into small chunks ('horse pieces') and boiled in open 'try' pots on deck to obtain the oil; once cooled the oil would be bailed into casks for stowage below. In sperm whales, the head was reserved for special treatment. It would be brought on board, usually in two pieces – the 'junk' (a fibrous mass from the lower part), and the 'case' (the upper part, containing highly valuable spermaceti). The liquid waxy oil in the case could literally be bailed out, and with oil from the junk would be stowed separately from the blubber oil. For right whales, the upper part of the head would also be hauled on board, but in that instance for removal of the baleen.

Figure 3.4 A bay whaling station in New Zealand, probably late 19th century. The major stages in oil production are illustrated – removing blubber from the carcass, carrying it up the beach, 'trying out' the oil in open boilers, and storing the oil in casks. Similar operations would have been seen on the Australian coast. Photo courtesy Bill Dawbin

As well as on the high seas, 'open boat' whaling was also undertaken from shore or in bays, on coastal species such as right whales and humpbacks. 'Shore whaling' involved an actual establishment, however crude and temporary, on shore, and 'bay whaling' involved a vessel that would remain moored in the bay and upon which operations were usually based. Both shore whaling and bay whaling were common in the early days of the Australian colonies, particularly in Tasmania, South Australia and Western Australia, as well as New Zealand. The late Dr Bill Dawbin, working out of the Australian Museum, Sydney, in the early 1980s, estimated that up to 50 shore stations were operating in south-eastern Australia in the late 1830s, although the exact number depended on whether temporary sites used for only one season were included. He records sites mainly on the Tasmanian south-east coast, in New South Wales near Eden (and further north), in Victoria near Portland and Port Fairy, and in South Australia at Sleaford Bay and between Port Lincoln and Encounter Bay, as well as on Kangaroo Island.

A shore station would usually be in a sheltered bay with a good water supply. The whalers would erect shelters and possibly a ramp for hauling up the blubber. A watch tower or convenient headland would serve as a lookout for passing whales. The whale boats and their gear were modelled on the open ocean pattern. Such activities tended not to last long, possible only for a few years until the local supply of whales disappeared, or the market for oil (or whalebone, or both) became depressed. Shore whaling for right whales was particularly short-lived, especially as it was largely aimed at the cows nursing their young close to shore. Within three or four years most of the adult females accustomed to visiting the bay every three years to give birth to their young might have been caught and the operation would fade away. An exception was at Twofold Bay, Eden, in New South Wales, where humpbacks were caught from the mid-1800s until around 1930. This whaling for humpbacks was uniquely characterised by the presence of killer whales, which apparently cooperated in the hunt.

In 'traditional', open boat whaling, only the slower-swimming species that floated when dead, that is, sperm and right whales, could be pursued with any real chance of success. Humpbacks were taken occasionally, but they were notoriously dangerous when attacked and were very much a second choice. The larger, also highly desirable species – blue whales, fin whales and sei whales – all swam too quickly for pursuit by oar or sail, moreover they sank when dead. That all changed with the advent of the fast catcher boat: powered by steam, and with a cannon in the bows firing an explosive-tipped harpoon, it had powerful springs to take the strain of the animal on the end of the line, and a compressor to pump air into the carcass before it had time to sink.

The Norwegian Svend Foyn developed this new system in the 1860s. It was used with considerable effect, at first in coastal whaling off northern Norway and elsewhere in the north-east Atlantic. From 1904, it was used in the Antarctic, beginning at South Georgia in the South Atlantic on the then copious stocks of blue, fin, humpback and right whales there. At first, whaling could only be conducted from land-based stations, or from ships moored in sheltered bays. But in the mid-1920s the next and final development, the factory ship, freed the operation to operate anywhere on the high seas. Its fleet of catcher boats would bring the carcasses to the factory where they could be hauled up through a stern slipway to be dismembered on deck, for processing in cookers below. Thus was possible the huge commercial open ocean slaughter that began in the late 1920s and only ended, when most southern stocks were virtually exhausted, in the mid 1980s.

Figure 3.5 Sperm whaling off Albany, Western Australia. The harpoon, tipped with an explosive grenade, has just been fired at the whale ahead of the catcher boat. Photo © Durant Hembree

Australia played an early role in Antarctic blue whaling when the pioneer whaling captain C.A. Larsen took the factory ship *Sir James Clark Ross* from Hobart into the Ross Sea in 1923–24. Eleven of the crew were Australians, among them the writer Alan Villiers, and the catch was largely blue whales.

Whale products

The main product throughout the centuries, both in traditional and modern whaling, has been oil, which is mostly obtained from the blubber. However oil is present throughout the body – as mesenteric fat, in the bones, and elsewhere. Its composition is quite different in baleen and toothed whales, so much so that great care was taken during processing to keep baleen and sperm whales separate.

Oil from baleen whales (commonly known as 'whale oil') is a relatively simple compound – a fatty acid glyceride. It was used unprocessed as fuel for lighting, but could also be 'hardened' to make margarine and lard, 'saponified' to make soap and glycerine, or 'polymerised' for use in varnishes and in the manufacture of linoleum and printing ink.

Sperm whale oil is, on the other hand, a liquid wax, which is much more complex chemically than baleen oil. And spermaceti, found particularly in the 'case' of the sperm whale head, is different again from sperm oil as such. In contact with air, sperm oil remains liquid and was used 'raw' for lighting and for lubricating machinery such as in steam engines. Spermaceti, unlike sperm oil itself, solidifies into a firm, white wax, and was used to produce the very finest candles, burning with an odourless, smokeless flame. As the major product of Yankee whaling, sperm oil was also used in cosmetics, for imparting a sheen. Later processes such as sulphurisation, saponification and distillation gave additional valuable products – for example for lubricating high speed machinery (as in automobile gear boxes), for scouring and bleaching cloth, and for softening leather. Sperm oil from Albany, Western Australia, was used, among other things, in the production of leather for the finest quality gloves.

The baleen of right whales, known generally as whalebone, was a most valuable product in traditional whaling. It could be extraordinarily profitable – the baleen from one whale might pay the expenses of the voyage – the rest was sheer profit. Its flexibility, springiness, and strength were used in making hoops for crinolines, corset stays, ribs for umbrellas, whip handles and fishing rods. It could also be split and woven to make chair seats, sieves, even nets. But its use died out as right whale stocks declined, and with the development of spring steel.

The modern industry went well beyond just using the blubber oil, spermaceti or baleen – it made use of the whole carcass. There was whale meat for human consumption – baleen whale meat being generally much more palatable than sperm meat. There were products such as meat extract (a kind of Bovril), meat and bone meal – used as animal feed and fertiliser. There were also 'solubles', a liquid concentrate remaining after processing the meat in the cookers, which was used as an additive in animal feed. There were various other specialised products, too – gelatin from the bones, vitamin A from the liver, and even hormones from the pituitary gland.

In addition to all the above, rather more exotic products such as ambergris and sperm whale teeth could be desirable (see Chapter 2).

Figure 3.6 The last whaling operation in Australia at Frenchman Bay, Albany, closed in 1978. From 1964 it caught solely sperm whales. Photo: Pat Baker © Western Australian Museum

The Australian whaling industry, operating after World War II, was based on the same products as described above for the modern industry. At Tangalooma and Byron Bay in the east, Point Cloates and Carnarvon in the west, as well as initially at Albany in the south-west, humpbacks were killed for their oil, 'solubles' and meat meal. From 1964, sperm whales were the only species taken at the Albany operation. They were killed for their oil, whale meat and solubles, and to a minor extent for ambergris and teeth. Most of the sperm oil was sold overseas. In 1977, a year before the operation closed down, 91 per cent was exported, as mixed sperm oil and spermaceti. Of the oil sold in Australia, most (over 85 per cent) was used in the lubrication and leather tanning industries. For sale within Australia, sperm oil was filtered, to remove the spermaceti. All spermaceti used in Australia was imported as the amount obtained from the local industry was too small to justify refining here. Of the meal and solubles, about 40 per cent was sold overseas, while the remainder was used mainly as additives to pig and poultry feed.

'Non-consumptive' uses

A popular development from the early 1960s was the establishment of aquaria presenting displays of live cetaceans. There was a major boom in such establishments in the 1970s but this was largely negated in the 1980s with growing pressure from animal rights lobbyists. At the same time, it is practical to keep only the smaller cetaceans, and no great whales, successfully in captivity, although two young gray whales were held for a while on the west coast of the USA. Zoos and oceanaria maintaining smaller cetaceans, including bottlenose dolphins, or even animals as large as killer whales, have undoubtedly played a major role in the public's understanding of cetaceans. But in Australia their establishment has been effectively banned since 1985. In that year, resulting from a Senate Select Committee review (the Georges Report), the government decided not to grant any further permits for the purpose. Currently only two such establishments exhibit dolphins in Australia: Sea World on the Gold Coast, Queensland, and the Pet Porpoise Pool at Coffs Harbour, New South Wales.

A major development involving the great whales has, however, proved extremely popular and is a major boost to local economies. Whalewatching is now enjoyed by many millions of people worldwide each year. Many programs are directed to dolphins and porpoises, but whalewatching began with gray whales off the coast of California, USA. In the winter of 1950, 10 000 people watched gray whales passing San Diego from a land-based lookout. In later years, the whalewatching took place also from boats. From gray whales the whalewatching industry spread to fin, minke and beluga whales in the St Lawrence River, Canada, and then to its major flowering from the mid 1970s on humpbacks, at first off New England and Hawaii. From the 1980s, whalewatching spread to other countries: sperm whales in New Zealand, right whales in Argentina and South Africa, sperm and minke whales in Norway, Bryde's and sperm whales in Japan, and blue, fin, humpback and minke whales in Iceland. In 1998 alone it was estimated that more than nine million people in 87 countries or overseas territories went whalewatching, generating more than $US 1 billion.

As with aquaria and oceanaria, whalewatching has encouraged public interest in and sympathy for marine mammals. Many, though not all, operations include an educational element, employing naturalists – often young biology students – to lecture and answer questions. Many assist research by maintaining sightings records and obtaining identification photos, thereby contributing to public support for marine mammal research.

In Australian waters whalewatching has concentrated mainly on humpbacks particularly off Perth, Western Australia, and in Hervey Bay, Queensland. Blue whales are watched in Geographe Bay, Western Australia,

Figure 3.7 Whalewatching on humpback whales off the coast near Perth, Western Australia. Photo: © Chris Burton

and there is boat-based watching of right whales and humpbacks at Albany, Western Australia, and land-based right whale watching at Fitzgerald River National Park, Western Australia; Head of Bight, South Australia; and Warrnambool, Victoria. In the Great Barrier Reef, a unique commercial operation allows people to 'swim' with dwarf minke whales in accessible lagoons during the winter.

These are just some examples of whalewatching in Australia. The various State governments provide more information and a comprehensive listing can be found on the website of the Department of the Environment, Water, Heritage and the Arts, Canberra. The industry operates under guidelines developed jointly in 2005 between the Australian Government and the States, and each State is in the process of implementing them. They deal with the operation of vessels and aircraft, as well as swimming and diving, near all cetaceans. The guidelines fall within a general framework of objectives and principles adopted by the International Whaling Commission based on advice from its Scientific Committee. In that context the Commission has mirrored the popular view that whalewatching is a form of sustainable use of cetacean resources.

4
BIOLOGY

No branch of zoology is so much involved as that which is entitled cetology.

William Scoresby, quoted by Thomas Beale in
The Natural History of the Sperm Whale, *1839.*

Toutes ces indications incomplete ne serve qu'a mettre les naturalistes á torture.

Baron Georges Cuvier (about cetacean biology),
quoted by Thomas Beale, as above. Freely translated as
'such incomplete information just gives naturalists nightmares'.

Thomas Beale, a surgeon who served on two British sperm whaling vessels in the early 1830s, gave one of the first, if not in fact the first, accurate accounts of the sperm whale's appearance, habits and general biology in his classic work, *The Natural History of the Sperm Whale*. Things have moved a long way since his time, but as is obvious in what follows, there are still very large gaps in our knowledge of even some of sperm and other whales' most basic biology, as exemplified by the frequent use of 'seems to', 'may be', 'apparently', 'believed to be', in what follows, which concentrates on Australia's great whales.

Evolution and special adaptations

Cetaceans are generally believed to have evolved from artiodactyls (even-toed ungulate land mammals) related to such present-day forms as cows and hippopotamuses, although exactly which is a matter for argument.[1] The order Cetacea has three suborders – Archaeoceti, Odontoceti and Mysticeti, of which only the last two survive today. The earliest cetaceans, the archaeocetes, appeared first about 50 million years ago. The odontocetes and mysticetes diverged from a common archaeocete ancestor between 25 and 35 million years ago. Some archaoecetes possessed feet, while some early mysticetes possessed both teeth and baleen, later losing the teeth (except as foetal remnants) but retaining the baleen.

Modern whales differ from their ancestral archaeocetes mainly in having a 'telescoped' skull. 'Telescoping' results from the migration of the nasal openings (to become the blowholes) to the top of the skull, but achieved in different ways in mysticetes and odontocetes; it leads to considerable extension of the upper and lower jaws and a well-developed 'rostrum' or beak. In addition, modern whales have a fixed elbow joint; in archaeocetes it is flexible and can be rotated, presumably an indication of their relatively recent terrestrial origin.

Modern toothed whales characteristically possess a single blowhole (although whether that is actually the case in sperm whales is controversial), a 'melon' – a fatty tissue lump forming the 'forehead' and thought to be used in focusing sound in echolocation – and an asymmetrical skull (where the right side is more developed than the left). The eight families comprise the sperm whales (two families), beaked whales, river dolphins (two families), narwhal/beluga, 'true' dolphins (including killer whales and pilot whales) and porpoises. Of those, the river dolphins occur only in South America or Asia, while the narwhal and beluga are solely arctic. In the context of this book, only the sperm whale, among the toothed whales, is a 'great whale'. Sperm whales possess a characteristic spermaceti organ, and functional teeth only in the lower jaw. In the fossil record there are some 20 or so sperm whale genera, in two subfamilies: one, now extinct,

---

1 A recent find by Dr Hans Thewissen of the Northeastern Ohio Universities of Medicine suggests the ancestor of cetaceans may have been a small deer-like animal *Indohyus*.

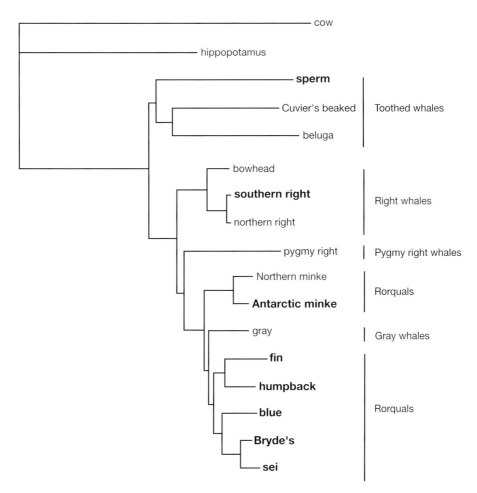

Figure 4.1 Phylogeny (evolutionary tree) of the whales, showing the relationships of the various taxa and their origin from terrestrial even-toed ungulates (cows and hippopotamuses). The great whales described in this book are shown in bold. Not included is *Balaenoptera omurai*, probably closer to blue whales than Bryde's whales. Modified from Rychel et al., 2004

had functional teeth in both jaws; the other, leading to present-day sperm whales, had rudimentary teeth in the upper jaw.

Modern baleen whales have lost their ancestral teeth and have developed very large body size, large heads and short necks. In addition to the very well-developed rostrum, the connection at the front of the lower jaw is loose and flexible. These adaptations provide for the long baleen plate row hanging from the upper jaw and allow the mouth to open very wide.

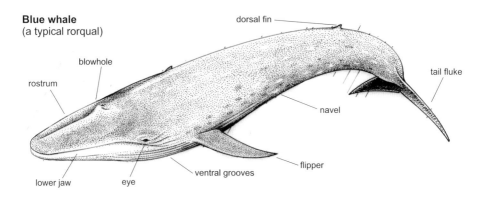

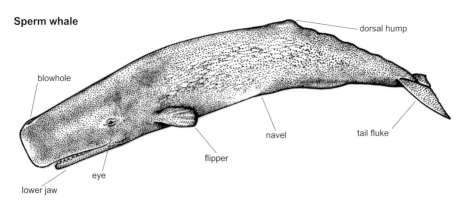

Figure 4.2 Some key parts of the whale.

Cetaceans have become extremely specialised for a life spent entirely in water. They have had to respond to the needs of swimming, diving and communication in an essentially buoyant, resistant, cold and salty

Figure 4.3 The head of a male sperm whale caught off Albany, Western Australia, shows the remarkable underslung lower jaw (mandible) with its two rows of functional teeth. When its mouth is closed, its teeth fit into sockets in its palate. The whale is lying on its left side; a front mandibular tooth has been removed to determine its age. Photo: Pat Baker © West Australian Museum

environment. Their bodies have become streamlined, almost spindle-shaped; the forelimbs are flattened as flippers, used for maintaining stability and steering; the hindlimbs are totally lost externally, although retained as internal pelvic remnants; horizontal tail flukes, with no skeletal support, have been developed for propulsion. The typical mammalian hair covering is lost, although a few bristly hairs can be found on the jaw and head in some species; the testes are withdrawn into the body cavity and the penis is retractable; the nipples are contained in slits. While whales have very small and inconspicuous earholes, they do not have external ears. A blubber layer has been developed for insulation; it also helps to streamline the body, and is important for buoyancy and as a major energy store. Internally, lung volume is reduced, and the lungs can be collapsed during diving (to avoid build-up of nitrogen and 'the bends'); the diaphragm is orientated diagonally and is longer and stronger than in other mammals to allow evacuation of a large amount of air in a short

Figure 4.4 The head of a fin whale taken off South Georgia, showing the array of baleen plates forming the sieving 'doormat' on their inner surface. The whale is lying on its back and the lower jaw has been removed.   Photo: John Bannister

time: 80 per cent of air in the lung may be exchanged at one time. The blood has an exceptionally high myoglobin content for increasing its oxygen-carrying capacity; heart rate can be reduced during diving; and breathing is under conscious control. In response to the salty environment the kidneys are lobed, each lobe corresponding to an individual kidney; there are very many more lobes for example in baleen whales (3000) than in land mammals with lobed kidneys such as cows (25–30). The sense of smell is reduced in baleen whales and absent in toothed whales. There are conflicting views on cetacean sound production: the larynx has become modified both in baleen and toothed whales (see Sound production, Chapter 4), and in sperm whales the spermaceti organ has been invoked, controversially, as playing the major role in production of the sperm whale's chief vocalisation – the click.

Life history

Cetaceans, and particularly the great whales, as long-lived animals with relatively delayed sexual maturity and relatively long reproductive interval, have evolved to maintain relatively stable population sizes at or near the environment's carrying capacity (usually denoted as K). They are therefore characterised as K strategists, in contrast, for example, with small, rapidly reproducing terrestrial mammals with large numbers of offspring (characterised as r strategists), or even other marine mammals such as sea otters or seals. One would therefore expect the great whales to be unable to react swiftly to major changes in their environment or to severe reduction in their numbers. Nevertheless some cetacean species are able to do so rather more effectively than others, as will be shown below.

Swimming and diving

Among the baleen whales, the rorquals are the fastest swimmers: sei whales have been recorded at around 35 knots (nearly 65 km per hour) in short bursts; fin and minke are also fast, at up to 20 knots (ca 37 km per hour). But the most powerful swimmers are blue whales, able to maintain more than 15 knots (28 km per hour) for several hours. On migration over long distances, humpbacks have been recorded at an average of just under three knots (a little over 5 km per hour). There are no long distance figures for southern right whales, but off coastal southern Australia Dr Stephen Burnell, of Sydney University, has recorded cow/calf pairs travelling at 1.5–2.3 knots (2.7–4.2 km per hour) in 24 hours. Sperm whales are also relatively slow-moving, swimming at just over two knots (around 4 km per hour).

Marine mammals have been described as 'commuting' to the depths. To do that they have to adapt to rapid changes as they move from the surface to several hundred, or even a few thousand, metres deep.

Baleen whales do not generally dive very deep, relying for food on prey found relatively close to the sea surface (see Table 4.1). The deepest diving baleen whale is the fin whale, which although recorded to a maximum depth of 500 metres usually dives 100–200 metres, for a maximum dive-time of 30 minutes. Humpbacks have been recorded to around 150 metres and a maximum dive-time of around 20 minutes. Blue whales have generally been believed to be fairly shallow divers, to no more than 100 metres, but for not more than around 30 minutes, although there is at least one echosounder record of a blue whale diving at around 400 metres in the Perth Canyon, Western Australia (see Figure 4.5). Right whales have been

Table 4.1 Swimming speeds and diving depths of the various species

Species	Swimming speed while feeding or on passage		Maximum swimming speed		Diving depth
	knots	km/hr	knots	km/hr	metres
Southern right whale	1–2	2–4	?8	?15	180
Blue whale	1.5–3	3–6	>15	>28	to ca 400
Fin whale	5–8	9–15	20	37	100–200
Sei whale	1.5–7	3–13	ca 35	ca 65	?1–200
Bryde's whale	1–4	2–7	11–13	20–25	300
Minke whale	1.5–7	3–13	20	37	?100–200
Humpback whale	1.5–3	3–6	15	28	150
Sperm whale	1–3	2–6	12+	22+	>1000 (males) 300–800 (females)

recorded down to about 180 metres and a maximum dive-time of 50 minutes. But the supreme diver is the sperm whale, recorded to depths of well over 1000 metres and maximum down-times of more than one hour.

Migration
Baleen whales undertake some of the longest migrations known of any mammal. Southern hemisphere humpbacks may cover as much as, or even more than, 50° of latitude either way between feeding and breeding grounds, a round trip of some 6000 nautical miles (11 000 km) (see also Chapter 5, Humpback whale). 'Australian' animals can move between feeding grounds south-west of Western Australia at 65°S and breeding grounds off the north-west coast at around 15°S. In the region south of New Zealand they may well travel from almost 70°S (in the Ross Sea, particularly from just south of the southern boundary of the Antarctic Circumpolar Current in water temperatures of 0°C) to around 17°S, near Fiji.

While southern blue and fin whales feed extensively in Antarctic waters in summer, their calving ground locations are not known. That they do migrate has been strongly inferred from winter catches of both species in lower latitudes, for instance off southern Africa, and to some extent confirmed by the (admittedly small number of) Australian records, for example of fin whales there, in winter, together with a lack of winter sightings on pre-World War II dedicated Antarctic cruises. Sei whale migrations are relatively diffuse and may vary greatly from year to year

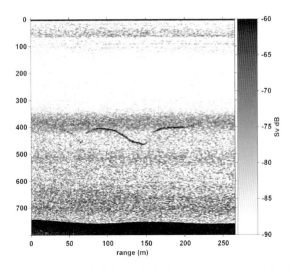

Figure 4.5 An echosonogram of a blue whale (presumed pygmy blue) diving at a depth of some 400 metres in the Perth Canyon, Western Australia, March 2003. The sigmoid trace between 393 and 465 metres depicts the whale. The base of the Leeuwin current is at 360 m with a dense layer down to 416 m. The seabed is at ca 750 m.   Courtesy of Rob McCauley, CMST, Curtin University and Curt Jenner, Centre for Whale Research, WA.

in response to environmental conditions. On the other hand, Bryde's whales hardly migrate at all; they must satisfy their reproductive and nutritional needs exclusively in tropical/temperate waters. The same presumably applies to pygmy blue whales, which may travel further south than Bryde's whales, to ca 55°S, but must also rely very largely on feeding outside Antarctic waters for their needs. Their irregular appearance in various localities from year to year, as in the Perth Canyon or the Bonney Upwelling, seems to be closely linked to the availability of food there. Even among migratory animals like humpbacks, it's possible that not all individuals migrate each year; genetic samples off both the east and west Australian coasts have been biased towards males, which with a similar bias in the catches has been taken to show that not all females migrate north every year. The matter is however controversial and arguments over possible bias in the samples, as well as in the catches, remain unresolved. Nevertheless, in both humpbacks and fin whales there is well-defined segregation of various classes of animals on migration. Immature humpbacks and females accompanied by yearling calves migrate north first, followed by adult males and non-pregnant adult females; pregnant females bring up the rear. On the southern migration the same pattern is followed, but with cows accompanied by their calves of the year travelling last. But not only is there segregation within some (possibly all) species, the species themselves arrive on and leave the feeding grounds at different times. At South Georgia in the South Atlantic, where they were once common, fin whales tended to leave the feeding grounds after blue whales but before sei whales, the latter having arrived there most recently.

By comparison with most baleen whales, the other great whale, the sperm whale, undertakes much less well-defined movements. It has been said that except for humans and killer whales, few animals on earth are as widely distributed as sperm whales – they can be found near the ice-edge in either hemisphere, and are common along the equator, particularly in the Pacific. There is however a very marked difference in distribution between females (and their young) and the older males. The females and young are found only in latitudes between 40° in each hemisphere (except for the North Pacific where they go as far as 50°N) and where sea surface temperatures are warmer than 15°C. The males on the other hand move toward the poles, with the larger and older males becoming more solitary in the coldest waters.

Social activity
Baleen whales rarely form large aggregations, except when feeding. On migration, most baleen whales travel in small groups. Larger groupings, as seen for example in oceanic dolphins, are said to be a response to predation; much bigger individual animals such as baleen whales will be subject to low predation pressure and can therefore afford to travel in smaller groups. Blue and fin whales are usually encountered singly or in groups of two to three, but feeding concentrations of up to 100 or more fin whales have been reported. Sei whales can also be found in large feeding groups, but on migration usually travel in groups of no more than six or seven. Minke whales also concentrate together on the feeding grounds but on migration usually travel as twos or threes. Sperm whales, on the other hand, as already described, can form large migrating herds. Usually only the larger, older, males are found as solitary animals, and then in colder waters, although as large 'socially mature' breeding males they return to warmer latitudes; they then roam independently between groups of females. They seem to spend only a short time actually in such schools, and when roaming seem generally to avoid other males. But as Dr Hal Whitehead, of Dalhousie University, Canada, concludes, sperm whales must occasionally fight, given the deep scars from other males' teeth often seen on the head and body.

The social behaviour of great whales has been studied most in coastally breeding humpbacks and right whales, and also in sperm whales. Male humpbacks compete for females by singing and fighting. The songs act as some kind of courtship display, but whether to attract females (the 'best' singer getting the 'best' female) or to repel other males is unknown. Humpback males can be very aggressive towards each other. They congregate near an adult female, fighting for position: they can lunge with ventral grooves extended, hit with their tail flukes, slap their flukes and tails, even omit streams of bubbles from the mouth. Individuals may be left raw and bleeding from the sharp barnacles on the skin. Right whales also congregate in what are known as 'surface active groups' (SAGs), but their constituent individuals are far less aggressive and larger numbers may interact – 11 or more animals have been recorded together. It has been suggested that the female may even initiate the behaviour. In the North Atlantic right whale, SAGs have been observed throughout the year, including outside the main breeding season, and at that time may represent more of a strictly social interaction. Indeed the right whale breeding regime is believed to involve sperm competition rather than direct competition between males for a female (see Chapter 5, Southern right whale).

Sound production
Evolving in a marine environment where sight and smell are less useful than sound, most cetaceans have lost a sense of smell and their eyesight is limited. They have to rely heavily on sound and hearing, with which they communicate and become aware of the world around them. But there are major differences between baleen and toothed whales in their ability both to produce and to hear sounds, and this is reflected in their anatomy. Baleen whales, like humans, have a larynx, but unlike humans (where in speech air is forced out of the larynx through the vocal cords into the open) baleen whales lack vocal cords. It seems that instead of forcing air out into the open, baleen whale sounds may be produced by recycling air through sinuses in the head. Toothed whales, on the other hand, while also lacking vocal cords, possess 'phonic lips' protruding into the nasal passage and past which air is pushed to produce sounds; it is then released into the open as a stream of bubbles (as in many dolphins) or recycled through a 'vestibular sac' just below the blowhole.

Typically, very low frequency communication sounds are produced by the two largest baleen whales, blue and fin, although in fin whales these are much shorter ('downsweeps' of only about one second) than in blue whales (typically 20 seconds long); both species produce calls made up of such sounds in long, regularly spaced patterns. While the calls may be related to courtship, in fin whales at least they can be heard year-round, leading to the suggestion that individuals must be advertising constantly. Fin whales have also been recorded making other, 'rumble', sounds particularly near ships, taken to indicate that the whale has been surprised. Variations on the long calls in blue whales can also be found. Indian Ocean blue whales produce 'songs' in the form of repeated sets of sounds which don't seem to occur in other blue whale populations. The Indian Ocean blue whales may belong to the pygmy blue subspecies – indeed nine different kinds of blue whale call have recently been described worldwide, presumably denoting different populations (see Chapter 5, Blue whale). It has been suggested that the loud, low frequency, calls of blue and fin whales, able to be heard over long distances of at least several hundred kilometres (and perhaps further in the right conditions), allow the formation of 'acoustic herds', where information can be exchanged, for example on the availability of food, or within which contact is maintained.

Other rorquals make characteristic calls. Minke whales consistently seem to make short (0.2–0.3 second) downsweeps; the dwarf minke has been recorded making a remarkable 'star wars call' (see Chapter 5, Minke whale). Bryde's whales moan, sei whales have been recorded calling in two very short phrases (each 0.5–0.8 seconds long). But humpbacks are acoustically the best known, and most studied whales, for their remarkable vocal repertoire. It includes 'songs', changes in which among Australian east coast humpbacks have been cited as an example of 'cultural evolution' (see Chapter 5, Humpback whale).

Right whales, in the southern hemisphere studied acoustically off Argentina by Dr Chris Clark of Cornell University, USA, produce 'up' calls about one second long, thought to be used in gathering individuals together. 'Down' calls are also heard, thought to be used in maintaining contact, for example between an adult female and her calf. Other noises include 'sweeps' and pulses, in addition to noises produced by the blow, by slapping flippers or flukes on the water, even by 'rattling' the baleen plates.

The largest toothed whale, the sperm whale, like other toothed whales, not only produces sounds for communication but also presumably for echolocation, a facility not available to baleen whales. The major sounds consist of sets of loud and structurally complex 'clicks'. Clicks can exist singly, in a short pattern (as 'codas'), or in long sequences ('creaks'), with different frequencies depending on the animal's sex – large males have lower frequency clicks than the (smaller) females and younger males. Codas appear to be for socialising, particularly among females in nursery groups. Most scientists believe that regular trains of clicks, produced at depth and at about one to two per second, are used in echolocating for food: accelerating series ('creaks') are thought to be produced as the animal closes on its prey. But that belief (proven for dolphins) has been challenged recently for sperm whales, given that such sounds seem unlikely to be able to detect the whale's commonest food – squid, which are usually less than one metre long and have acoustic properties the same as seawater. It may be that such clicks are reserved for catching fish whose bony skeletons and swim bladders provide excellent acoustic targets; sperm whales may catch squid visually and fish acoustically. It has also been suggested that clicks are important on the breeding grounds where lone, breeding, males may 'gauge' each other, and, conversely, females may 'gauge' such males. In producing such sounds, and/or in diving, the role of the spermaceti organ has been a matter of some debate (see Chapter 5, Sperm whale).

In the Perth Canyon and elsewhere off Western Australia, Dr Rob McCauley of Curtin University, Western Australia, has been undertaking a long-term acoustic study, based largely on 'sea noise loggers' moored on the ocean bed, but including some drifting. Among many thousands of hours' results, several different whale noises have been recorded, including:

- from pygmy blue whales – heard very commonly in summer
- from possible Antarctic blue whales – although not positively identified, they resemble Antarctic blue whale calls in structure
- from probable fin whales – generally resembling fin whale calls described in the literature
- possible minke whale calls
- humpback songs.

The pygmy blue calls seemed to have a common structure and several variants, including possible social calls. Pygmy blue whale calls generally last two minutes and are spaced about 80 seconds apart. Calls are more frequent (2.2 times) at night than during the day. Calling animals seem to be spread fairly evenly over a large range, up to perhaps several hundreds of kilometres away along the shelf edge, where extremely good transmission can be obtained; a logger on the bottom at 450 metres depth received calls from at least 50 km away, and probably much further. Long-term results from loggers elsewhere, including off Cape Leeuwin, Exmouth and the north-west shelf edge, as well as in the Perth Canyon, have been used to suggest a southerly pattern of pygmy blue whale movement along the Western Australia coast, with an influx of animals from the north passing Exmouth in late spring (November), arriving later off Perth and Cape Leeuwin with a peak in March–May and a northbound pulse mid-year. Sea noise loggers set by Dr Jason Gedamke of the Australian Antarctic Division have also recorded pygmy blue whale calls along the Polar Front (formerly known as the Antarctic Convergence) south of Tasmania at 55°S in the southern summer. Dr McCauley has extensive records of pygmy blue whale calls from western Victoria and northern Tasmanian waters in late summer into autumn. These records suggest the animals roam widely across southern Australia and at least south to the Polar Front (that is, to around 55°S).

The Antarctic blue whale call, which differs from the pygmy blue whale call heard in summer along southern and western Australia, has been recorded across southern Australia during winter from Bass Strait to Cape Leeuwin. In one set of recordings Antarctic blue whales seem to have remained in Bass Strait just north of Tasmania for several weeks.

A major acoustic monitoring program studying changes in song patterns of humpback whales is being undertaken by Dr Michael Noad and colleagues at the University of Queensland. The Humpback Whale Acoustic Research Collaboration (HARC) is a major collaborative project with participants from Australia, the USA and the UK. Two large field experiments were conducted in 2003 and 2004 at Peregian Beach, Queensland; data analyses are ongoing. The aim is to learn more about how sounds from other species and the environment affect the behaviour of east coast migrating humpback whales. Given current concerns about the effects of anthropogenic noise on marine mammals (see Chapter 7), HARC is part of a global effort to obtain data for the development of improved models and has been recognised worldwide as a template for multi-disciplinary collaborative work on whales.

In addition to HARC, the University of Queensland and the Defence Science and Technology Organisation (DSTO) are involved in the long-term monitoring of humpback whale songs off the Australian east coast. The songs change yearly (see Chapter 5, Humpback whale) but the factors driving the changes are not understood. Most of the recordings for this important long-term data set have been collected at Point Lookout and Peregian Beach, Queensland, and Cape Byron, New South Wales.

Habitat

Between them the great whales occupy most available marine habitats. As already noted, sperm whales are very widely distributed, from the polar ice edges to the equator, and together the various baleen whale species cover much the same range. Among the latter, fin and sei whales are probably the most oceanic though fin whales penetrate much further into colder seas than sei in summer. Blue whales are also highly oceanic, but can on occasions be found close inshore. Most if not all summer records of blue whales in southern hemisphere low latitudes are likely to be pygmy blues (Antarctic blues should at that time be feeding in the Antarctic); at such times pygmy blues may be found in coastal areas where food concentrations (generally small krill species) occur – Australian cases in point are in the Perth Canyon area off Western Australia and the Bonney Upwelling off south-western Victoria. The numbers appearing in such places can however be quite irregular, even though there tends to be a 'peak period' of occurrence (in the Perth Canyon generally February–March), presumably linked to the irregular occurrence of their prey, in turn affected by prevailing oceanic conditions. On the other hand the regular appearance of blue whales very close to the coast in Geographe Bay, on the south-west coast of Western Australia in October and November, seems not to be linked with feeding at all but more as some kind of staging post on a southern migration. It must be significant that a blue whale successfully satellite-tagged there was recorded six weeks later some 500 km south of Esperance, Western Australia, near the Subtropical Front, and that it remained there – a known feeding ground for pygmy blues – for two weeks. It may also be significant that some animals recorded in Geographe Bay at that time may well, judging from their size, be Antarctic blues, and if so, as suggested by Chris Burton, of Western Whale Research, who has been undertaking a long-term program of blue whale observations in Geographe Bay, are presumably on their southward migration to Antarctic feeding grounds.

The most coastal rorquals are humpbacks, with their long migrations between low latitude breeding grounds and high latitude feeding grounds. In the southern hemisphere their routes include the east and west coasts of the three continents – including off Australia. Minke whales, though the smallest rorquals, go as far south as any, up to and even into the ice in summer, but like blue and fin whales, their winter breeding grounds are unknown. Where the dwarf minke whale, relatively common off Australia in winter – and, as already noted, well known in the Great Barrier Reef as the subject of 'swim' programs – go in summer is a mystery. They are rarely found in cold waters, and must, like pygmy blues, rely for their food on ephemeral concentrations of warm water krill species and other small prey. On the other hand, the most localised baleen whale, Bryde's whale, is restricted entirely to tropical and warmer seas and seems not to undertake long migrations, although the two forms, offshore and inshore (see Chapter 5, Bryde's whale) differ in their movements: the inshore form may be coastal throughout the year, whereas the offshore form appears and disappears seasonally, presumably in association with the occurrence of its main food, shoaling fish.

While sperm whales can be found in any ocean, and in the tropics at least at any time of the year, their distribution is far from even: whalers relied on a knowledge of their concentration areas ('grounds') for their catches. Generally, sperm whales' favourite habitat is deep water in areas of high productivity, usually resulting from upwelling; off Australia they are most commonly found on or close to the continental slope, as off Albany, Western Australia, south of Kangaroo Island, South Australia, and off the west coast of Tasmania. These localities are associated with submarine canyons dissecting a considerable rise over only a short distance from very deep water – off Albany, Western Australia, the seabed rises from 5000 to 200 metres over only few kilometres. There seem to be generalised sperm whale movements towards the poles in summer, but near the equator there is little evidence of seasonal movement. Off Albany there was a very well-marked progression westwards through a narrow strip along the shelf edge, animals travelling steadily at 2–3 knots (just over 4 km per hour), and passing through the whaling area in 24 hours. Most animals were in 'nursery' groups (adult females and young) separated from aggregations of 'bachelor' males (averaging around 10 animals per group). In some parts of the world female sperm whales seem to have home ranges about 1000 km across, but this depends upon the abundance of food. Large males are usually found only in deep water, but usually where productivity is high, as along continental shelf edges.

Food and feeding

Despite including the largest living animals, baleen whales rely for food on some of the smallest. They are strictly carnivorous, 'filter feeding' on zooplankton or small fish. Toothed whales however, including the sperm whale, while still strictly carnivorous, feed quite differently, generally capturing their prey individually.

All baleen whales possess a filter based on baleen plates borne transversely on the upper jaw, but two rather different feeding systems have developed – 'skimming', and 'lunging' (or 'gulping'). In each a large volume of water containing food organisms is filtered through the baleen's fine inner fringing hairs, combining to form a kind of filtering doormat. 'Skimming' feeding is practised largely by right whales. The highly arched jaw and massively enlarged fleshy lower lips accommodate very long and narrow baleen plates (up to nearly three metres long in right whales, even more – to four metres – in the Arctic bowhead). A gap in the front between the rows allows the whale to scoop up a great quantity of water as it swims slowly forward through a prey school. 'Lunging' or 'gulping' feeding occurs in the rorquals, with their much shorter, triangular baleen plates. Here the animal opens its mouth extremely wide in combination with opening the lower jaw almost to 90° from the body axis, depressing its tongue and expanding its ventral grooves. It then lunges into a school of prey, gulping in great quantities of water and straining the food out through the fringing 'doormat'. The two systems allow, on the one hand, the slow swimming right whales to concentrate their rather sparse prey over a period and, on the other, the fast swimming rorquals to obtain large amounts of their more concentrated prey in a shorter time.

Baleen whales typically feed on swarming plankton, generally euphausiids or copepods. Apart from the warm water-loving Bryde's whales, dwarf minkes and pygmy blues, feeding is mostly in colder waters. Unlike many toothed whales, including the sperm whale, baleen whales don't dive very deep to feed, usually to only within about 100 metres of the surface – although an example has already been noted of a blue whale in the Perth Canyon sonically detected diving at around 400 metres. Baleen plate structure, particularly the fringing hairs, mirrors to some extent the food taken, or for Antarctic krill, the size classes. Thus blue, fin, humpback, sei and right have inner fringes of increasing fineness; similarity in that character between sei and right whales has led to the possibility that they may compete for the same food – copepods (but see below). Bryde's whales, feeding largely on fish, have altogether much coarser, bristly, baleen fringes. It's been calculated that a large baleen whale needs about four per cent of its body weight in food per day during the feeding season, while a pregnant female may need to increase its body weight by up to 65 per cent to meet the enormous demands of pregnancy and lactation. This just goes to show how efficient the feeding system is (and how nutritious the food) for such an increase to be possible in only a few months' feeding.

Sperm whales feed very largely on deep sea squid, although males tend to eat larger individuals than do the females. In some waters, such as in the North Atlantic near Iceland, the main diet is fish. Only 1.6 per cent of sperm whales examined in the commercial catch off Albany, Western Australia, 1964–66, had been feeding on deep sea fish, and over 70 per cent had been feeding just prior to capture off the edge of the continental shelf, from the evidence of squid or fish flesh present in the stomachs examined. But how sperm whales actually catch their prey has been the subject of much debate. Studies by Prof. Malcolm Clarke of the Marine Biology Association, UK, showed that the diet of sperm whales caught near the Azores in the North Atlantic was only 23 per cent fast-swimming large squid; mostly the diet was smaller, neutrally buoyant luminescent species, with about 1000 items taken per day. The received wisdom has, until at least recently, been that sperm whales, like dolphins and bats, use echolocation to locate their prey, but it has been suggested that their 'clicks' are unlikely to be able to detect objects less than one metre in diameter (see earlier this chapter, under Sound Production). Also squid tissue has acoustic properties too similar to seawater to reflect sound. It may be that sight, and the property of squid to bioluminesce when stimulated, play a much larger part in sperm whale feeding behaviour than previously imagined.

The question of competition between species is contentious, and has been raised in the cetacean context in a number of ways. Dr R M Laws, of the British Antarctic Survey, calculated that the great reduction of baleen whales by whaling in the Southern Ocean to around one-third of their original numbers and one-sixth in biomass must have left a large surplus of food, some 150 million tonnes, available for other consumers, such as seals, penguins and fish. In response there may well have been increases in growth rates, earlier maturity, and higher pregnancy rates in some baleen whale species. But the evidence is equivocal. Whether individuals actually compete on the same prey, at the same time and in the same area, is questionable. The observed increases since the 1980s in southern right whales and humpbacks, and more recently in blue whales (and also possibly fin whales), suggest that competition is unlikely at least where, as in the Antarctic, food supplies are abundant. But the question has been raised both in relation to the second phase of Japanese special permit catches in the Antarctic (JARPA II, see Chapter 6), as well as to the possible effects of climate change (see Chapter 7).

Predators and parasites

Humans have been the most notable and dangerous cetacean predator, but apart from them, killer whales are major predators on several species. Killer whales have been reported attacking blue and sei whales; minke whales have been identified as a major item of their diet in the Antarctic. Killer whale tooth marks are often reported on humpback whale tail flukes. They have been seen to attack sperm whale calves: the adult females at that time form a protective circle round the victim with their tails outward (the well-known 'marguerite' or 'daisy' pattern); a variation on that theme can occur, where the adults line up tightly together and face the attacker. Sharks can prey on humpback and right whale calves on the breeding grounds; anecdotally, false killer whales have been reported attacking humpback calves.

An apparently unique cause of concern has been described from the major right whale calving ground at Peninsula Valdes, Argentina, by Victoria Rowntree, of the University of Utah, USA. There, kelp gulls have developed the habit of gouging skin and blubber from the backs mainly of 'logging' adult right whales, causing adverse reactions and, where the adults are suckling their calves, possibly affecting calf development. So far nothing of that kind seems to have been reported for right whales elsewhere, including Australia.

External parasites include whale lice (cyamid crustaceans, probably more properly termed ecto-commensals) and barnacles. Both are commonest on the slower moving species, humpbacks and right whales. In the latter, light-coloured cyamids aggregating on the head callosities (see Chapter 5, Southern right whale) have facilitated individual photoidentification studies. The highly specialised copepod *Penella* occurs anchored in the blubber of fin and sei whales in warmer waters; its presence on whales at places such as South Georgia in the South Atlantic has been taken as evidence of recent movement from lower latitudes. The cold water diatom *Cocconeis ceticola* can form a brownish-yellow film on the skin of blue (where it led to the early common name 'sulphur bottom') and other whales in the Antarctic. It takes about a month to form, and so has been used to indicate how long an animal might have been there. The origin of small scoop-like bites on the skin of baleen whales was unknown until it was discovered they were caused by the bites of the small warm water shark *Isistius* (also known as the cookie-cutter shark). Their overlapping healing scars on southern hemisphere sei whales can impart a 'galvanised iron'-like sheen to the body.

Growth and reproduction

Among the great whales, most baleen whales generally follow a similar pattern of growth and reproduction. Mating occurs in warm waters in winter, followed by a gestation period of around 11 months. The cow may then suckle its young for seven months or more, including the calf's first journey to the feeding grounds where it takes its first solid food. The cow may then spend a year or more 'resting' before the next pregnancy. Most adults become sexually mature at around 5–10 years, and can live for 50 years or more. The largest baleen whale, the blue whale, can live probably to 80 or 90 years and reproduce every two to three years. The smallest, the minke whale, may live to 50 years. They can also become pregnant again immediately after birth. That has also been recorded in humpbacks, allowing depleted stocks to recover rapidly.

Right whales follow a similar pattern, but lactation may last up to a year, and while females can reproduce successfully from about six years, the calving interval is a relatively regular three years. Sperm whales on the other hand, as might be expected, are quite different. Gestation may last up to nearly 16 months, followed by a very long lactation period. Calves may take solid food after a year but suckling may continue for several years. Births may occur at intervals of four to five years.

The absolute age of an individual may be determined if it can be identified first as a calf and then later in life, as for photo-identified animals such as right whales (using the head callosity pattern). Otherwise age has to be estimated using tissues from dead animals, either stranded or taken in whaling. In rorquals, people have used counts of ridges in the baleen plates, or of growth layers in the tympanic bullae, but both have limitations: baleen plates wear rapidly at the tip, and bulla counts seem to be incomplete. Layers not subject to loss through wear are, however available in the rorqual 'earplug' – a waxy formation in the external auditory meatus where layers accumulate throughout life. For toothed whales, similar growth layers are found in the teeth, although care has to be taken to obtain an unworn, or if not, the least worn, tooth in the row. Earplugs are not easy to collect and can be fragile, but when bisected and polished on the cut surface, they display a series of light and dark layers. Each pair of layers (known as a Growth Layer Group, GLG) is taken to represent one year's growth. There has been some controversy in the past over exactly what is being counted, and how many layers are formed each year, but based on a study of fin whale earplugs from the Antarctic in summer and warm waters in winter, Dr Howard Roe of the UK Institute of Oceanographic

Sciences demonstrated conclusively that one dark and one light layer represented one year's growth, and that has been generally accepted for rorquals as a whole.

In sperm whales the most unworn tooth, usually the front mandibular, is bisected and etched, usually in formic acid. The GLGs can then be shadowed, by rubbing with pencil or photographically, and counted. The method of tooth etching and photography was developed for sperm whale age determination by Jeanne Bow and Charles Purday at the CSIRO Fisheries and Oceanography laboratory, then in the mid 1960s in Sydney. A study by Dr Karen Evans at the University of Tasmania showed that individual workers may vary considerably in their interpretation of GLGs: the method requires care and experience, and consensus readings are necessary for its success. In addition, direct evidence for the annual deposition of growth layers in sperm whale teeth is still lacking.

There has been recent interest in the possibility that telomeres (nucleoprotein structures at the ends of chromosomes that degrade during life) may be very useful in age determination. They have the considerable advantage that they can be obtained from only a small sample of skin, and therefore non-invasively, but so far it has not proved possible to calibrate the rate of degradation and thus provide a reliable estimate of the animal's age.

5
SPECIES ACCOUNTS

The Indian Sea breedeth the most and the biggest fishes that are: among which the Whales and Whirlepooles called Balaenae …
Holland's Pliny, quoted in Moby-Dick, *1851.*

This chapter gives details of the eight species of 'great whale' (as defined for the purposes of this book) found in Australian waters. Information is provided on their general appearance, distribution, movements, food and feeding, breeding, and current status – including each species' most recent IUCN 'red list' category, and its status under the *Australian Environment Protection and Biodiversity Conservation (EPBC) Act 1999.*

Blue whale
Balaenoptera musculus

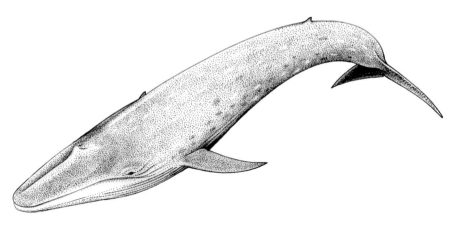

The largest animals ever, blue whales are huge, growing to around 30 metres in length and a maximum weight of more than 130 tonnes. Despite their bulk, like other rorquals they are remarkably streamlined. The body is a mottled bluish-grey, the head is broad and flat-topped (rather U-shaped when seen from above), the flippers are slender and pointed, there is a small 'nub' of a dorsal fin well back on the body, and the tail flukes are triangular. The single blow is very tall and columnar. Underwater, blue whales appear distinctively light blue-green. The short baleen plates are all black, with coarse inner fringes. At sea the very tall blow, huge size and long back, appearing well before the small dorsal fin, are characteristic.

In the southern hemisphere two subspecies are recognised: the Antarctic – sometimes known as the 'true' – blue (*B. m. intermedia*) and the pygmy blue (*B. m. brevicauda*). As well as being generally smaller than Antarctic blues, growing only to about 25 metres, pygmy blues, as their scientific name indicates, are relatively shorter in the tail, that is behind the dorsal fin. While it is difficult to distinguish between the two subspecies at sea, Dr Hidehiro Kato of Tokyo University of Marine Science and Technology has shown that Antarctic blues are 'torpedo'-shaped when seen from above or behind, while pygmy blues, with a relatively broader head and shorter tail, are 'tadpole'-shaped. Genetic distinctions have been described at the population level between Antarctic and pygmy blues, but no distinctive diagnostic genetic marker has yet been found for individual animals.

An alternative, now archaic, name for the Antarctic blue, 'sulphur-bottom', refers to the yellowish film of diatoms (*Cocconeis*) acquired ventrally after a prolonged spell in Antarctic waters.

The species occurs worldwide, in all oceans except the Arctic, but not in some regional areas such as the Mediterranean. Northern hemisphere animals (as in other rorquals) are generally smaller than in the southern hemisphere. As for most baleen whales, the females are slightly larger than the males. Blue whales can live for 80–90 years.

Blue whales are rather solitary animals, usually found singly or in pairs. They are very powerful swimmers, able to maintain speeds of over 15 knots (over 28 km per hour) over long distances. Over five seasons of observations in the Perth Canyon, Western Australia, feeding blue whales (presumed pygmy blues) were observed by Curt and Micheline Jenner of the Centre for Whale Research and colleagues diving for an average of nine minutes, with intervals at the surface of around three minutes, during which they averaged nine 'blows'.

Blue whales make long and very low frequency sounds, at around 17–20 Hz, lower than humans can detect, but extremely powerful. They also make sounds at higher frequencies, up to several hundred Hz. At 188 decibels the low frequency sounds have been described as louder than a jumbo jet engine only a few metres away – one of the loudest and lowest made by any animal. The calls can be heard over large distances – under ideal conditions possibly across ocean basins; the low frequencies are obviously well-designed for communication through water between animals far apart from each other. There are distinct differences between the sounds made by Antarctic and pygmy blues, but just to confuse matters, the sounds recorded by different pygmy blue populations may be as different as between them and Antarctic blues. A recent study has delineated nine different blue whale regional songs worldwide, varying in complexity from simple (including animals from the Antarctic) to more complex (including off Chile and New Zealand) and very much more complex – and longer – than the others (from the Indian Ocean, including off Western Australia).

Distribution and movements

Like most rorquals, blue whales feed extensively in cold waters during summer, and are generally presumed to undertake long migrations to warmer waters in winter for breeding. The exact locations of the breeding grounds are unknown but that they do migrate seems evident from winter catches off southern continents, for example, off southern Africa. Pygmy blues are generally restricted to waters north of around 55°S, and are more localised in their distribution. They seem to undertake less regular movements, relying more opportunistically on localised food concentrations; their stronghold is the Indian Ocean. An apparently separate population, found only in the northern Indian Ocean, may be a separate subspecies, as may be a localised population off western South America.

There are records of blue whales, both Antarctic and pygmy, from all Australian states. But by comparison with the two more coastal great whales, right and humpback, only small numbers were killed in Australian waters, presumably because of their generally more offshore distribution and lack of aggregation during migration. There are few confirmed records of Antarctic blue whales in Australian waters, although a skeleton of a physically immature animal in the Western Australian Museum, from Busselton, Western Australia, in 1898, at around 24 metres in length has been assumed to be an Antarctic blue. During Australian coastal humpback whaling, 1954–63, 22 blue whales, all pygmy blues, were taken. In the 1960s, Soviet whalers took pygmy blue whales along much of the southern and western coasts, from the Great Australian Bight to at least Shark Bay in WA. Recent Australian blue whale strandings have been identified as pygmy blues; they included two mature female pygmy blue whales one year apart in Princess Royal Harbour, Albany, Western Australia, in 1973 and 1974. Of 15 blue whales recorded to 45°S on a whale sighting cruise off south-western Australia in February–March 1993, one was identified as a 'true' (Antarctic) blue, the rest as pygmy blues. Up to two per day were recorded in December 1995 on a dedicated blue whale sightings cruise in a small area some 25 nautical miles off Rottnest Island, in the area of the Perth Canyon, Western Australia, and almost five per day in an area close to Portland, Victoria, later the same month. Other key Australian localities include Bass Strait, south-east Victoria, south-west New South Wales, and off east and west Tasmania. Apart from those already mentioned, strandings have been recorded in South Australia, Victoria, Tasmania and Queensland.

Recent studies in the Perth Canyon by a consortium of scientists from Curtin University (Drs Rob McCauley, Susan Rennie and Chandra Salgado Kent), the Centre for Whale Research (Curt and Micheline Jenner), Western Whale Research (Chris Burton) and the Western Australian Museum, have recorded sightings peaking there in late summer (February–March). Studies by Dr Peter Gill and Margie Morrice of Australocetus Research and Deakin University, Victoria, in the Bonney Upwelling, near Portland, Victoria, have recorded blue whales there through summer and into autumn, attracted by well-defined krill concentrations. Sightings by Chris Burton in Geographe Bay, south-western Western Australia, close to the coast in September–December each year, have been presumed to be mainly pygmy blues, although from their size, some Antarctic blues may also be there then. Acoustic data from the same area remain unidentified but may be of Antarctic blue whales. Curt and Micheline Jenner have recorded sightings, both from aerial and vessel surveys, of presumed pygmy blue whales (based on body shape as seen from the air – 'tadpole' rather than 'torpedo'), off north-west of Western Australia, along the continental shelf break near North West Cape, and north of the Dampier Archipelago. In general, it seems reasonable to assume that most blue whales recorded off Australia in summer will be pygmy blues, although in winter they may be either subspecies.

Satellite marking, using tags developed by Dr Nick Gales at the Australian Antarctic Division, Hobart, together with photo-identification studies, has recorded a small number of blue whale movements. One animal moved south to 45°S from Geographe Bay, WA; another had moved between Portland, Victoria, and the Perth Canyon, and there are indications of movement northwards from the Perth Canyon. Aerial surveys by Peter Gill and his colleagues between Bass Strait and the Great Australian Bight, mostly in the Bonney Upwelling, have progressively expanded known (presumed pygmy) blue whale feeding habitat along Australia's south-east coast. One satellite-tagged animal showed movement from the Bonney Upwelling area to the south as far as the northern boundary of the Subtropical Front, around 40°S. The emerging picture is thus of a more or less continuous blue whale (presumed pygmy blue) population around the Australian coast from Bass Strait, moving as far south as at least 45°S, and up the west coast possibly towards Indonesia.

Feeding and food

Like other rorquals, especially fin whales and humpbacks, blue whales are 'lunge' feeders. They swim into large prey swarms, gulping in huge quantities of water and food and straining the food through the matted inner fringes of the baleen plates. Again as in other rorquals, they are able to do this through a combination of a large expandable mouth, a widely opening lower jaw (almost to 90° from the body axis), and numerous extensible ventral grooves (see Chapter 4).

Antarctic blue whales feed almost exclusively on Antarctic krill, *Euphausia superba*. Pygmy blues feed on smaller euphausiids. In the Perth Canyon they have been recorded feeding on swarms of *Euphausia recurva*; in the Bonney Upwelling the main food is *Nyctiphanes australis*. They feed both at the surface and deeper: an acoustic record in the Perth Canyon shows a blue whale diving at 400 metres (see Chapter 4).

Breeding

Mating and calving occur in winter, with a 10- to 11-month gestation period. Weaning probably occurs after about seven months, the calf taking solid food during its first visit to the Antarctic. Sexual maturity can probably first occur at around five years old, but may range to 10 years. Calving probably occurs at around two to three year intervals. The exact location of blue whale breeding grounds is unknown.

Status

Antarctic blue whales were reduced to a minute fraction of their original numbers by greatly excessive whaling, at first in the 1920s when 'modern' pelagic (open sea) whaling began and again after World War II. The original population, estimated at some 240 000 animals, may have been reduced to fewer than 1000 by the 1960s. There were considerable fears that with such low numbers the Antarctic blue might actually become extinct, but recent sighting surveys have shown a small but significant overall increase, to around 2000, at an encouraging increase rate of about seven per cent. Even so the current Antarctic population is still less than three per cent of its original population size.

No estimates are available for pygmy blue numbers overall in the southern hemisphere, but nearly 13 000 were killed in the southern Indian Ocean over only a few years in the 1960s, suggesting the population there was at least originally that size.

No population estimates of Antarctic blues are available for Australian waters, but some information on pygmy blues is available from recent surveys in the Perth Canyon area off Perth, WA. The numbers visiting that area in late summer (peaking in February–March) are only small, and very variable; they probably represent a small fraction of a more widely dispersed population (see above) numbering perhaps in the hundreds. In a series of aerial surveys over the period 2000–2006, the estimate in the survey area (some 50 x 40 nautical miles centred some 60 nautical miles off the coast) ranged between only 4 (in 2002) and 47 (in 2004); the average over the seven year period was 30. The estimates, undertaken by Dr Sharon Hedley of St Andrews University, Scotland, use the number of animals seen from the aircraft, adjusted to allow for animals missed during diving. No estimate is yet available of the numbers in the Bonney Upwelling area from year to year, but the impression is that they may at times (as in December 1995) be more concentrated there than off Perth.

Given their still very low numbers, Antarctic blues have been listed as Critically Endangered by IUCN. Under the *Environment Protection and Biodiversity Conservation Act 1999* (EPBC Act; see Chapter 6), blue whales are classified as Endangered.

Fin whale
Balaenoptera physalus

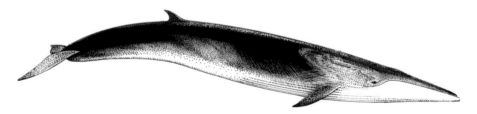

The second largest baleen whale, the fin whale, with its long slender body, up to around 25 metres long and a weight of some 90 tonnes, is well adapted to its life as the 'greyhound of the oceans', although the sei whale is faster. The fin whale's head is slender and has a distinct median dorsal ridge. As in blue whales, the flippers are short and narrow, the tail flukes are triangular and the numerous ventral grooves reach back to the navel. Behind the tall hooked dorsal fin the back is distinctly ridged, leading to the variant, but now unfashionable, name 'razorback'. At sea the 'inverted cone' blow is impressive, up to 6 metres high, and visible over a great distance; the 'wheeling' motion of the body at the surface, where the dorsal fin appears shortly after the tall blow, is characteristic. The fin whale rarely raises its tail clear above the water ('flukes up') before diving.

Fin whales are generally dark grey on the back and sides, and white underneath. The undersides of the flippers and tail are white. Just behind the head there is often a greyish-white chevron, its apex orientated forward. But the most distinctive feature is the weirdly asymmetrical colour of the lower jaw, which is white on the right and dark on the left. This is mirrored by a group of white baleen plates on the right side of the mouth. The rest of the baleen plates, on both sides, are black, and all have rather coarse yellowish-white inner fringes.

As in most baleen whales, the females are slightly larger than the males. They can live for over 90 years. Fin whales are not normally gregarious, usually being found singly, or in small groups of up to seven animals; they may gather together in larger groups for feeding. Normal swimming speed is 5–8 knots (9–15 km per hour); in short bursts fin whales can reach around 20 knots (about 37 km per hour). They rarely breach, and usually perform shallow dives at between three and 10 minute intervals. The depth of their dives ranges from 100 to 200 metres, although they have been recorded to

500 metres. There is distinct segregation on migration, with pregnant females travelling first, followed by adult males; lactating females and juveniles bring up the rear.

Fin whale sounds consist of simple low frequency moans and grunts and high frequency pulses. The moans are very loud and can be heard over at least tens, possibly hundreds, of kilometres.

Distribution and movements

Fin whales occur worldwide, including the Mediterranean, but in the southern hemisphere they tend not to go as far south as blue, humpback or minke whales, and are not generally found close to the ice edge; they occur mainly between 40°S–60°S in the southern Indian and Atlantic oceans and 50°S–65°S in the South Pacific. They are highly migratory and undertake extensive journeys between warm water breeding grounds and cold water feeding grounds. Off South Georgia in the South Atlantic, where they were once very common, they tended to leave the feeding grounds after blue whales but before sei whales. Off Australia their migration paths seem to be generally oceanic, individuals only infrequently being recorded near the coast. They were rarely taken at Australian land stations: only three were recorded caught between 1949 and 1963. They do strand occasionally – one notable animal washed ashore dead near a major Perth bathing beach in July 1996. There are no obvious key localities for the species in Australian coastal waters, although fin whale sounds have been detected, for example off the Perth Canyon, Western Australia.

Feeding and food

Like blue whales and humpbacks, fin whales feed by 'lunging' into dense swarms of their prey. As in other rorquals, they are able to do this through a combination of a large mouth, a widely opening lower jaw (almost to 90° from the body axis), and numerous extensible ventral grooves (see Chapter 4). Swimming into large prey swarms, they gulp in huge quantities of water and food and strain it through the matted inner fringes of the baleen plates.

In the northern hemisphere, fin whales are known to feed on small schooling fish such as herring and capelin, but possibly only opportunistically when these are abundant. In the southern hemisphere, and like blues and humpbacks, fin whales feed almost exclusively on Antarctic krill, *Euphausia superba*, during the summer.

Breeding

As in blues and humpbacks, mating and calving occur in winter, with a 10- to 11-month gestation period. Weaning may occur after some six to eight months. Sexual maturity is reached at between six and 10 years old. Calving occurs at around two to three year intervals.

Again, as for blue whales, no well-defined breeding grounds have been identified but there were extensive catches in winter off southern Africa (off Durban and near Cape Town) as well as off Angola, Congo and Mozambique; and also off Brazil, Chile and Peru. Reports of animals chasing each other in pairs or in groups of three in autumn may be related to mating activity.

Status

Fin whales were second only to blue whales in importance in the Antarctic whaling industry. A very large number, over 700 000, was taken between 1905 and 1976. The inevitable result was very considerable reduction, to perhaps as low as 15 000 in the early 1980s. Direct evidence of the reduction comes from the considerable decline in catch and sighting rates, of 89–97 per cent, off South Africa over the period 1954–1975, reported by Professor Peter Best, of the Mammal Research Institute, University of Pretoria, working from the Cape Town Museum. There is happily now some evidence of an overall southern hemisphere increase in numbers since the late 1970s. There are no estimates specifically for the Australian region, although relatively large numbers were reported in summer 2005–2006 during the Japanese special permit program (JARPA, see Chapter 6) in the Antarctic: 188 schools, comprising 748 animals, were sighted between 80°E and 135°E, that is, in Area IV, south of Western Australia. There have been other recent Antarctic sightings of more than 100 animals off the Ross Sea. Fin whales are listed as Endangered by IUCN and as Vulnerable under the Australian EPBC Act.

Sei whale
Balaenoptera borealis

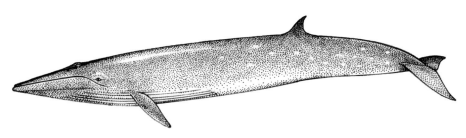

At around 20 metres in length and weighing around 20 tonnes, the sei whale is the third largest baleen whale after blue and fin whales. Distinguishing sei whales at sea from fin whales (and particularly from the sei whale's close relative, the Bryde's whale) can be difficult. As in other rorquals, the flippers are small and pointed, they and the triangular tail flukes are relatively smaller than in blue and fin whales. The dorsal fin is relatively large and markedly hooked.

The baleen plates of sei whales are generally all black, though some may be half white near the front. The inner fringes are white and very fine, almost silky in texture.

The top of the head has a single median ridge which distinguishes the species from Bryde's whale, which usually has two additional head ridges. Unlike blue, fin and Bryde's whales, the ventral grooves all end well forward of the navel. The body is dark grey dorsally and on the hind part of the ventral surface. The body often has a 'galvanised iron' appearance, caused by numerous healing lamprey or small ('cookie-cutter') shark bites. The flippers and tail flukes are dark underneath. Unlike fin whales, sei whales have grey lips. As in other rorquals, females are slightly larger than the males. A sei whale can live for at least 65 years.

The common name (often pronounced 'sigh', but more correctly 'say') comes from the Norwegian word for the schooling coalfish, or *saithe*, with which they often associate off the coast of northern Norway.

Sei whales usually form small groups of less than half a dozen individuals, but may aggregate in larger numbers on the feeding grounds. They have been recorded as blowing several times at 20–30 second intervals, followed by rather shallow dives for about 15 minutes or longer. They are the fastest baleen whale swimmers, having been recorded at around 35 knots (nearly 65 km per hour).

Little is known of sei's use of sound. 1.5–3.5 kHz calls have been recorded in the North Atlantic, while calls around 400 Hz were recorded near feeding sei whales in Antarctic waters.

Distribution and movements
Like most other rorquals, sei whales are cosmopolitan and oceanic in their distribution. Like blue and fin whales they undertake long migrations between summer feeding grounds and winter breeding grounds; there were catches in winter off Brazil, Peru, Angola, Congo, and western and eastern South Africa. They are however more irregular in their movements than other large baleen whales. In the southern hemisphere they do not travel as far towards the south pole as blue, fin and minke whales; in summer in the South Atlantic and Indian oceans they occur mainly in the region 40°–50°S, and 45°–60°S in the South Pacific.

Sei whales are not frequently recorded from Australian waters. Only four were caught from Australian whaling stations between 1958 and 1963. There are no obvious key Australian localities. Even though they were the most frequently reported baleen whale in sightings by whalers off Albany, Western Australia, during sperm whaling, they were not distinguished in the records from the more warm-water Bryde's whale. They are sometimes seen feeding in the Bonney Upwelling off south-east Australia between November and April. They have been recorded north of the Subtropical Front at 37°S, about 300 nautical miles (550 km) south-west of Ceduna, South Australia, and along the Front south of Tasmania.

Feeding and food
Like blue whales and humpbacks, sei whales can feed by 'lunging' into dense swarms of their prey, but they are also 'skimmers', swimming through prey schools with the mouth open and filtering their food like a right whale. Worldwide they are also fairly generalist feeders, taking small fish, squid, krill and copepods. In the Antarctic their main food is krill, but in less polar southern waters, for example around the Subtropical Front (and as in the North Pacific and North Atlantic), they can take copepods. That has led to speculation that there may be competition between sei and the other southern copepod feeder, the right whale, although that actually seems unlikely (see Chapter 4).

Breeding
As in blues and humpbacks, mating and calving occur in winter, with a gestation period that lasts 10 to 11 months. Weaning occurs after some six

to eight months, on the feeding grounds. Sexual maturity is reached at between six and 10 years old. Calving is at around two to three year intervals. Again, as for blue and fin whales, no well-defined breeding grounds have been identified.

Status

Although like other rorquals they could be caught only after the development of modern whaling, and then at first in the North Atlantic (hence the Norwegian name), sei whales were unimportant in southern whaling until the larger species, blue and fin, had been heavily depleted. Sei were taken in large numbers in the 1960s, when some 110 000 were killed between 1960 and 1970 (20 000 in 1964 alone). Most catching was in summer by pelagic fleets south of 40°S, but there was also winter catching from land stations in Brazil, Peru and South Africa, as well as from Chile (though there the catches were not distinguished from Bryde's whales). The inevitable result was very severe depletion. The southern population is thought to have originally numbered around 100 000 and may have been reduced to fewer than 10 000, but there have been no recent estimates, and sightings are rarely reported. Unlike blue, fin, humpback and right whales, there is no recent evidence of any recovery in numbers since whaling ceased: one difficulty is that sei whales occur further north than the main areas recently surveyed (south of 60°S). Sei whales worldwide have been listed as Endangered by IUCN and are designated Vulnerable under the EPBC Act.

Bryde's whale
Balaenoptera edeni

Bryde's whales are the least known of the larger baleen whales and can easily be confused with sei whales at sea. Although shorter than sei, growing only to at most 15 metres in length, their general body shape is very like that of sei whales. The main distinguishing feature is the usual presence of three parallel longitudinal ridges on the head, compared with the sei whale's one.

Bryde's whales are generally dark above and lighter below. Unlike sei whales, their ventral grooves extend to the navel, and the baleen plates, while dark in colour, generally have distinctly coarse, lighter-coloured, inner fringes.

The common name, sometimes pronounced 'brides' or 'breeders' but more correctly 'brooders', comes from Johan Bryde, a Norwegian pioneer of South African whaling in the early 1900s The scientific name *'edeni'* commemorates the British High Commissioner in Burma where the type specimen originated.

Not only can Bryde's whales be confused with sei whales, but Bryde's whale taxonomy is itself confused. Another animal, *Balaenoptera brydei* (named after the whaling pioneer), was described from specimens taken off South Africa, but subsequently accepted as the same species as *B. edeni*.

Top: Ambergris from the Western Australian Museum collection (numbers M11222, M46299). The complete specimen on the right was found on the beach near Perth in 1995; it weighs about 300 grams. The other two show the typical internal concentric structure. Photograph by Patrick Baker © WA Museum

Bottom: Examples of scrimshaw. The four engraved sperm whale teeth and the fid (made of sperm whale jawbone, and used in splicing rope) to the left and rear centre are from the traditional or 'open boat' whaling era; the teeth were engraved with a sharp implement and usually highlighted with lampblack. The pair at rear bears figures typically copied from magazine illustrations. The single tooth left foreground has been identified as the work of 'Foster of Grimsby', ca 1840. The rest are 'modern': the three liqueur 'glasses' (turned from sperm teeth) are from Albany, Western Australia, late 1970s. The sperm tooth on the right is of Soviet origin and dated 1979; it was engraved using a 'hot wire'. From a private collection. Photograph by Patrick Baker © WA Museum

Top: An Antarctic blue whale 'blowing' at the surface. Note the patterning of light spots, the long back, and the flattened, 'boat-shaped', but fairly slender head.
Photograph by Paul Ensor © IWC

Bottom: A blue whale surfacing in the Perth Canyon, Western Australia. Compared with the Antarctic blue (top), the more rounded and broader head is diagnostic of a pygmy blue. Photograph by Curt Jenner © CWR

Top: A blue whale swimming slowly at the surface in Geographe Bay, Western Australia. Note the small dorsal fin. Blue whales approach unusually close to shore there. Photograph by Chris Burton © Western Whale Research

Bottom: A fin whale in the Antarctic, just beginning to submerge. Note the relatively large dorsal fin. Photograph by Paula Olson © IWC

Top: Two sei whales at the surface in the Antarctic. The dorsal fin is taller and more curved than in the fin whale. Photograph by Paul Ensor © IWC

Bottom: A Bryde's whale beside a dense shoal of schooling fish, just off Cape Cuvier, north of Carnarvon, Western Australia, in June 1993. The three parallel head ridges on the pointed head are diagnostic. Photograph by Curt Jenner © CWR

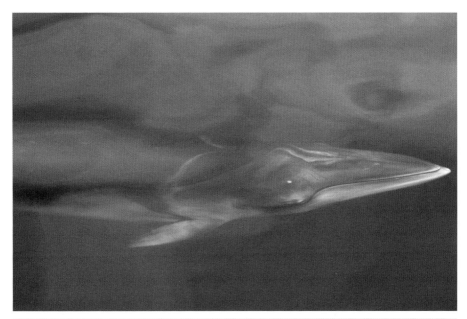

Top: An Antarctic minke just below the surface. The very pointed head is typical of minke whales. Antarctic minkes have no well-defined white patch on the flipper, by comparison with dwarf minkes (bottom) and northern hemisphere common minke whales.   Photograph by Paul Ensor © IWC

Bottom: A dwarf minke in the Great Barrier Reef, Queensland. The very striking black and white pattern is unique among baleen whales.   Photograph by Alastair Birtles
© James Cook University Minke Whale Project

Top: Two humpbacks off the Western Australian coast. The foreground animal is 'humping' its back as it dives; the background animal has a distinctly 'stepped' base to its dorsal fin. Photograph by Curt Jenner © CWR

Bottom: A southern right whale at the surface close to the surf line, near Augusta, Western Australia. The broad back, 'neck', and callosities on the head are typical.
Photograph by Chris Burton © Western Whale Research

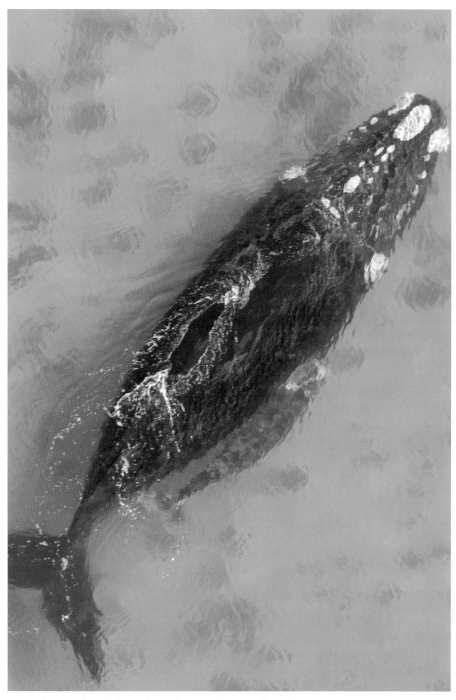

A southern right whale and newborn calf, east of Israelite Bay, Western Australia.
Photograph by Andrew Halsall © WAM

Top: A group of sperm whales at the surface off the Western Australian coast. The rounded 'hump', in place of the dorsal fin, is typical.   Photograph by Curt Jenner © CWR

Bottom: An old, lone, male sperm whale in the Antarctic. The single blowhole is very obvious at the front left of the head.   Photograph by Paul Ensor © IWC

Currently one species is recognised, but it has several forms, at least one of which may be a separate species. A 'regular' form has two distinct sub-forms – offshore and inshore. The situation has not been helped because the location of the type specimen of *edeni* was uncertain until recently and its genetic make-up has yet to be determined. A further similar but smaller species, Omura's whale, *B. omurai*, named after the distinguished Japanese scientist the late Dr Hideo Omura, was described in 2003, and recently accepted following genetic analysis, but it is not closely related to Bryde's whales. It is shorter than *B edeni*, growing to only 11 to 12 metres, and has only one median head ridge. A specimen in the South Australian Museum has been identified morphologically as *B. omurai*, but its genetic make-up has also yet to be determined.

Distribution and movements

Bryde's whales are unique among the larger baleen whales, being the least migratory rorquals, and restricted to tropical and temperate waters, from the equator to around 35°S.

Bryde's whales have been recorded from all Australian states except the Northern Territory, particularly in coastal localities such as the Abrolhos Islands and north of Shark Bay, Western Australia, and off Queensland. Given their generally warm-water preference, they are less likely to be found off the south coast.

There have been doubts about some Australian specimens' exact identity. Most Indian Ocean animals seem to be the large 'regular' form, but among eight individuals recorded by Dr Graham Chittleborough of CSIRO at Australian whaling stations from 1958, three animals taken off Western Australia and two from the east coast seemed to be intermediate between *B. edeni* and the sei whale, *B. borealis* (their ventral grooves, for example, reached short of the navel – the animals may even have been *B. omurai*), whereas three records from Victoria and another from Western Australia were of typical *B. edeni*.

Feeding and food

Given the coarse structure of their baleen and their warm-water habitat, Bryde's whales prefer larger food than schooling plankton, feeding mainly on pelagic schooling fish such as pilchards, anchovies and sardines. They seem however to be opportunistic feeders, and will take euphausiids, copepods, cephalopods and even pelagic crabs. The offshore form seems to feed more on euphausiids than the inshore form, which relies more on shoaling fish. There was a remarkable instance of Bryde's whales feeding on an extensive anchovy shoal close to the cliffs at Cape Cuvier, north of Shark Bay, Western Australia, in July 1993. The whales' behaviour, while spectacular, was not necessarily unusual – it just happened to be in a place where it could be recorded easily. Graham Chittleborough found that several of the Bryde's whales caught close to the coast off Carnarvon, Western Australia, in 1959 had stomachs containing large quantities of anchovies.

Breeding
Mating has been recorded throughout the year in the inshore form but during winter in the offshore form, with gestation lasting 11 to 12 months. Females may become sexually mature at around seven years and calving seems to occur mainly at two year intervals; weaning seems to occur at about six months.

Status
Before the 1970s, Bryde's and sei whales were considered together in official catch records. Although various attempts were made in the late 1970s and early 1980s by the International Whaling Commission's Scientific Committee to estimate Bryde's whale numbers in various regions of the southern hemisphere, no satisfactory estimates were obtained and they have not been reassessed since. Off Australia their occurrence is considered generally sparse, but since most records are from strandings this may represent an underestimate. The fact that six specimens of *B. omurai* originally came from deep water near the Solomon Islands, and two from near Cocos Island in the Indian Ocean, suggests it may be relatively more common in Australian waters than previously thought.

Minke whale
Balaenoptera bonaerensis/acutorostrata

Two species of minke whale are currently recognised: the northern hemisphere common minke, *Balaenoptera acutorostrata*, and the Antarctic minke, *Balaenoptera bonaerensis*. Additionally, there is a dwarf form, found only in the southern hemisphere and closely related to the common minke. Although clearly documented since 1985 it has yet to be formally described and named scientifically. Both the Antarctic minke and the dwarf form are found off Australia.

Minke whales are the smallest of the rorquals. The Antarctic minke grows to about 10 metres long and a weight of about 10 tonnes, while the dwarf minke grows to around seven metres. The head is narrow and pointed, with a distinct single median ridge. The dorsal fin is relatively tall and hooked, and situated further forward on the body than in other rorquals. The back is black and the belly and underside of the flippers white. Like the northern hemisphere common minke, the flipper of the dwarf minke has a distinct white patch on its upper surface; however this patch is usually absent in the Antarctic minke. The baleen plates are mostly black or dark grey, but a few on the left hand side and up to a third of those on the right at the front can be creamy white. Dwarf minke colouration is the most striking of any baleen whale: the combination of a dark throat patch, white shoulder region and white base to the flipper is diagnostic.

The name 'minke' is said to have originated when an inexperienced Norwegian whaler called Meincke misidentified a small whale as a blue whale – thereafter small rorquals were all 'minkie's whales'.

Minke whales usually occur singly or in groups of two or three, although on the feeding grounds, like other rorquals, they may aggregate together. In some areas they are well known for their 'ship-seeking' behaviour, where individuals approach slow-moving or stationary vessels. They are fast swimmers, with a maximum speed of some 20 knots (37 km per hour). Common minkes, closely related to dwarf minkes, have been recorded off Norway with large individual variation in surfacing rates, blowing at between 33 and 72 times per hour. Dwarf minkes, which usually occur alone or in pairs, regularly approach anchored charter boats and may approach snorkellers and divers. The maximum age for a minke whale may be around 50 years.

Minkes are said to produce a variety of sounds, including 'sweeps', grunts, whistles and even 'clanging bells'. Dwarf minkes have a distinctive, pulsed 'Star Wars' call that may represent a reproductive display in males.

Distribution and movements

Minke whales occur worldwide and undertake extensive migrations between lower latitude breeding grounds and polar feeding grounds, but less predictably than other rorquals: some, for example in the North Pacific, may not migrate at all. Antarctic minkes migrate further south than most rorquals except blue whales; they can be found in open water within the ice, and were the basis for the Antarctic 'pat the whale club'. Dwarf minkes have been recorded from Australia and New Zealand, South Africa and South America. They seem generally not to migrate to the Antarctic, although there have been occasional records as far south as 65°S. Many Australian records of minkes are of the dwarf form, from as far north as at least 12°S on the east coast and around 14°S on the west coast; they are well-known in the northern part of the Great Barrier Reef between 14° and 16°S in winter. Most records from subantarctic waters have been from December to March while off Australia they are predominantly from March to October; it is not known whether these are the same animals. Scattered records, together with the presence of fresh scars from the deep-water 'cookie-cutter' shark *Isistius*, suggest that they also spend time in the deeper waters of the Coral Sea.

Within the northern Great Barrier Reef dwarf minkes are usually encountered close to the shelf edge in water less than 50 metres deep, where they readily approach vessels and form the basis of a popular 'swim with' whales tourist attraction. 'Swim' programs have been established on humpback whales, dolphins, even whale sharks in various parts of the world, the latter for example off Ningaloo Reef, north-western Australia. A well-established program involving dwarf minke whales has operated within the Great Barrier Reef since the mid-1990s. Divers hold on to a line run out from a drifting vessel, and are approached by individual minke whales. The diver's behaviour is thus under control – indeed the whale's approach is under the whale's control, and disturbance to the whale is minimised.

Feeding and food

Like other rorquals, especially blues and fin, minkes are 'lunge' feeders. They swim into large prey swarms, gulping in large quantities of water and food and straining the food through the matted inner fringes of the baleen plates. Again, as in other rorquals, they are able to do this through a combination of a large mouth, a widely opening lower jaw (almost to 90° from the body axis), and numerous extensible ventral grooves (see Chapter 4).

Antarctic minkes feed predominantly on Antarctic krill, *Euphausia superba*, and may also take some smaller euphausiid species; near the ice edge they may feed on ice krill, *Euphausia crystallorophias*. A dwarf minke caught at 58°S had been feeding on myctophids (small 'lantern' fish, found at the surface at night), others at 60–61°S had been taking fish, and one at 65°S euphausiids. It is not known what these mainly lower latitude animals feed on there though the assumption is that they must, like pygmy blue whales, take advantage of local concentrations of small euphausiid species. Long-term studies by the late Dr Peter Arnold of the Museum of Tropical Queensland and Dr Alastair Birtles of James Cook University, Queensland, have shown that individual whales may return to the same area over several years, leading to the suggestion that there may be discrete, localised Australian populations of dwarf minke whales.

Breeding

Little is known about reproduction in minke whales generally. In some populations the pregnancy rate in adult females has been recorded as more than 90 per cent so they can give birth every year. Calving occurs in winter after a gestation period of around 10 months. The lack of records of calves accompanying cows in higher latitudes suggests that, at least in Antarctic minkes, weaning takes place before or just at arrival on the feeding grounds. Sexual maturity seems to occur at about seven or eight years.

Status

Although well known to 'modern' whalers in the southern hemisphere, minkes were considered too small to be of any commercial interest until the 1970s. By that time the larger species, blue, fin and sei, had been so reduced in number that minkes became the main target. But controversy has surrounded estimates of their numbers, in particular whether they were increasing before the 1970s (possibly through reduction of the other baleen whales), how many there actually were in the 1970s, and what their numbers are now. Circumpolar sightings surveys between 1982 and 1988 gave an Antarctic minke estimate of 760 000, but subsequent surveys indicated lower abundance, by perhaps as much as 60 per cent, for the period 1991 to 2004. That decline has not been generally accepted, and the matter has yet to be resolved, although there seems no doubt that the total southern hemisphere population size is currently in the several hundreds of thousands. South of Australia, based on results from Japanese special permit (JARPA) catches (see Chapter 6), at least two separate biological stocks appear to be present on the Antarctic feeding grounds, an Eastern Indian Ocean stock and a Western South Pacific stock, but not corresponding to the current Antarctic areas IV and V – that is, south of Australia between 70°E–130°E and 130°E–170°W respectively. There is some evidence for a mixing zone between the two at around 165°E, but there may be further subdivisions. Further analysis of the existing samples is required to clarify the position.

No population estimates are available for the dwarf form, and it has not been subject to significant exploitation. Of over 1700 minkes killed in the Antarctic between 1987 and 1993, only 16 were the dwarf form.

Humpback whale
Megaptera novaeangliae

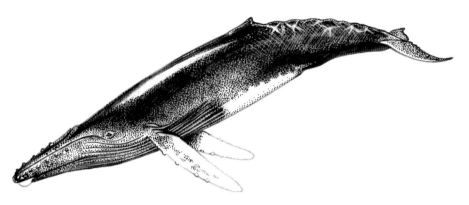

Compared with most other baleen whales humpbacks are rather short and stocky, growing to around 16 metres in length and a weight of some 45 tonnes. They can live for around 50 years. The name comes from the way they 'hump' their backs when they dive, emphasised by the dorsal fin, about two-thirds of the way back on the body. The fin can vary from a rather small rounded hump to a well-defined hook, but may have a fatty step or ledge at its base which emphasises the hump when the animal 'rounds out' as it dives.

The most distinctive feature of humpbacks is their long oar-like flippers. In most whales the flippers are short and paddle-shaped and used for steering or maintaining stability in the water. In humpbacks the flippers are remarkably long, up to about one-third of the body length, and mobile; they may be used to 'row' the animal through the water, and are sometimes waved about at the surface. Indeed the scientific name means 'large wing from New England', referring both to the flippers and to the origin of the first described specimen.

Humpbacks are distinctive in several other ways. The flippers have knobs on the leading edge; the head and lower jaw bear numerous tubercles, each usually the source of a single short hair or bristle; the broad tail flukes are scalloped on the trailing edge. The ventral grooves are wide and relatively few, and extend to the navel. The basic body colour is black dorsally and white ventrally; the flippers and tail flukes are white underneath. The greyish-black baleen plates generally have lighter-coloured, moderately coarse, inner fringes. Humpbacks are among the most heavily parasitised whales, with accumulations of barnacles, whale lice and diatoms on the skin. In males, a mass of barnacles on the chin can be used aggressively towards other males. Prof. Lars Walløe has recorded (see Chapter 2) the early Norwegian name *skeljung* – an animal 'covered by shells'.

Humpbacks are renowned as 'singers'. They have developed a sophisticated song – a complex set of notes in phrases, repeated continuously for 30 minutes or more. Singing seems to be restricted to males, and is presumably used in attracting females. At any one time, generally in the breeding season, all the males will be singing essentially the same song, but over the years it will generally change, though remaining distinctive for that particular population. In addition to their 'songs', humpbacks also make other, 'social', sounds – especially on the breeding grounds – and feeding calls.

Humpbacks are among the most highly active whales. They are well known for their spectacular behaviour, which includes breaching (leaping almost clear of the water), 'spy-hopping', 'lob-tailing', and waving the flippers in the air. They commonly raise their tails clear of the water ('fluke up') just before diving.

On migration they travel mainly in small groups, often only as singles or pairs, though larger groups may be encountered – usually no more than six or seven animals, occasionally up to 10. Off the Western Australian coast smaller groups (of one or two animals) have been recorded diving for some five to six minutes followed by about a further two minutes at the surface. Larger, more interactive groups dive for less time (about two and a half minutes) and spend relatively more time (about one and half minutes) at the surface. On migration over long distances they have been recorded as averaging around three nautical miles (5–6 km) per hour, although they can reach higher speeds, around 15 knots (28 km per hour). As in some other whales, particularly minke, young humpbacks are very curious and often approach slow-moving vessels, including whalewatching ones.

Distribution and movements
Humpbacks occur worldwide. Apart from one Breeding Stock in the Arabian Sea, all are highly migratory, undertaking long movements between summer feeding grounds and winter breeding grounds. At a round trip of over 6000 nautical miles (around 11 000 km), these are among the longest-known migrations of any mammal. A recent report details humpbacks travelling in the eastern South Pacific from the Antarctic and

Figure 5.1 Two humpback whales 'spy-hopping' at the surface. Photo © Chris Burton

crossing the equator to Costa Rica, a single journey of more than 4700 nautical miles (8600 km).

In the southern hemisphere, breeding tends to take place close to continental coasts or oceanic islands, with feeding in Antarctic waters. Currently at least seven southern hemisphere breeding populations are recognised, one each on either side of the three continents and another in the central South Pacific.

In Australian waters there are two breeding populations, one off the west coast, the other off the east coast (currently designated Breeding Stocks 'D' and 'E' respectively by the Scientific Committee of the International Whaling Commission). Breeding Stock D is relatively well-defined: mating and calving take place off the north-west coast in winter, with a major breeding ground identified by Curt and Micheline Jenner, of the Centre for Whale Research, off the Kimberley between Montgomery Shoal and Camden Sound, at around 15°30′S. Feeding occurs in the Antarctic in summer mainly between 80°E and 120°E, over most of the designated 'Antarctic Area IV', east of the Kerguelen Plateau.

The situation off eastern Australia is more complex. A subpopulation of Breeding Stock E migrates close to the coast and is believed to breed in the Great Barrier Reef and possibly in the Coral Sea near Chesterfield Reef, at around 19°S. Two additional subpopulations breed further east, that is, around New Caledonia and Fiji/Tonga, migrating through the Tasman Sea and past New Zealand. Breeding Stock E feeds in the Antarctic generally between 130°E and 170°W, over most of the designated 'Antarctic Area V'. Genetic studies, particularly by Dr Scott Baker and colleagues at the University of Auckland, New Zealand, have shown that the subpopulations of Breeding Stock E, that is animals breeding on the Australian east coast and further east, are more closely related to each other than they are to animals from Breeding Stock D. Some mixing occurs on the feeding grounds, varying over the years, between animals from Breeding Stocks D and E, though probably only a small amount, and possibly largely confined to the males. Indeed, a fascinating example of 'cultural evolution' was discovered by Dr Michael Noad of the University of Queensland and colleagues when typical east coast humpback song switched to a new, west coast version over a very short period, only three years, between 1995 and 1998, it seems as a result of the influence off a few male singers from the west coast. At the same time there is also some mixing, mainly on the feeding grounds, though currently to an unknown extent, between animals from Breeding Stock E and even further east, that is, in the central Pacific (currently designated Breeding Stock F): the exact relationships of animals from Breeding Stocks E and F have still to be unravelled.

Feeding and food
Southern hemisphere humpbacks feed mainly in Antarctic waters almost exclusively on krill, *Euphausia superba*. There is some evidence that they may feed on fish and plankton swarms in warmer waters – in the northern hemisphere they often feed on small shoaling fish – and during the southern migration off New Zealand they have been recorded as feeding on the larval form of the crustacean *Munida gregaria*. However, catches in the subtropics off north-western and eastern Australia in the 1950s showed almost no signs of local feeding, and Graham Chittleborough, working on the Australian whale catches in the 1950s and early 1960s, concluded that while they may occasionally ingest some plankton during the winter migration, the total food intake is negligible for at least four months each year.

Humpbacks are 'lunge feeders' – they swim rapidly into a prey school to gulp in large quantities of water and prey. As in the other rorquals and as opposed to the balaenids (right whales) they do this through a combination of a large mouth, a widely opening lower jaw (almost to 90°

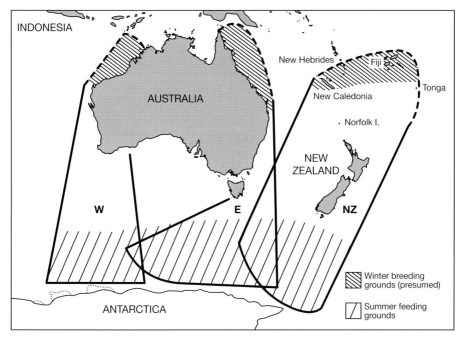

Figure 5.2 Distribution of the two main humpback Breeding Stocks (BS) in the Australasian region. Separate populations (BS D and E) winter off the west and east coasts and feed in the Antarctic in summer (here indicated as zones W, E and NZ), where there is some mingling. BS E is divided into populations wintering on and close to the Australian east coast (BS E(i)), and in the New Caledonia/Fiji/Tonga region (BS E (ii)). The latter is currently further divided into Breeding Stocks E (ii) 1 – wintering off New Caledonia – and E (ii) 2 – wintering off Tonga. Some animals, particularly those feeding in the New Zealand zone, may winter further east within Oceania. Modified from Bannister (1995), after Dawbin, 1966

from the body axis), and numerous extensible ventral grooves (see Chapter 4). But humpbacks have also developed another method, 'bubble feeding'. Although it is not often observed in the southern hemisphere, it involves the animal swimming in a circle emitting a ring of bubbles to encircle the prey, then swimming upwards through the prey with the usual lunging behaviour. That would not be expected in Australian waters, well north of the main feeding grounds, but there is at least one report of a bubble net associated with humpbacks in Shark Bay, Western Australia.

Breeding
Mating and calving occur in winter, with a 10- to 11-month gestation period. Weaning may occur after about six months, the calf taking solid food during

its first visit to the Antarctic, but there may be some suckling by the mother for longer, perhaps up to a year after birth. The age at sexual maturity has been a matter of some discussion recently. From animals of known age (using growth layer counts in the earplug, see Chapter 4) Graham Chittleborough determined that the mean age at puberty was four to five years in both sexes, but he based this on tracings of the baleen plates and his view that two earplug growth layers were formed each year. The validity of baleen plate tracings has since been questioned; it has also been established that in other baleen whales, particularly fin whales, only one earplug growth layer is formed per year (see Chapter 4). If this were the case for humpbacks, they would then mature much later, at around 10 years. Long-term photo-identification studies might be expected to clinch the matter, but photographs of animals on their breeding grounds can't resolve the problem, because the tail patterns in new born calves are not stable enough to be reliably recognised in later life. To complicate matters, in places where photo-identification can be used, as on humpback feeding grounds in the North Pacific and North Atlantic (the calves by then having grown sufficiently for their fluke patterns to stabilise), equally diverse ages at sexual maturity have been found, at around five years in the North Atlantic and more than 10 in the North Pacific. It seems that in any event there are likely to be differences in such values between populations, presumably caused by different environmental conditions, including availability of prey. Other influencing factors may include whaling history (where under intensive whaling late-reproducing animals would be less likely to live long enough to reproduce), or even migration length (where a longer migration could result in a shorter feeding season). Calving, on the other hand, for which there are good unequivocal data, usually occurs at around two to three year intervals, although one year intervals have been recorded.

Humpback males compete aggressively for females: highly interactive groups are often observed on and near the breeding grounds, with the males recorded as head-butting, ramming and tail-slashing. A major study by the late Dr Bill Dawbin of the University of Sydney, using all the available pre- and post-World War II catch data, showed marked segregation on migration. Among those travelling north, younger, recently weaned animals and mature females at the end of lactation travel first, followed by adult males, with pregnant females in the rear. On the southern migration, those first to arrive are the first to depart: recently impregnated animals leave first; cows accompanied by their newborn calves travel south last.

Figure 5.3 The underside of the tail fluke is widely used to identify individual humpback whales. There are individual differences in colour pattern and the scalloping on the hind border, as well as barnacle scars and killer whale tooth marks. Large catalogues exist of animals photographed off the Australian east and west coasts. Photograph © Doug Coughran

Status

Worldwide, humpback numbers were drastically reduced by commercial whaling, particularly 'modern' whaling in the late 19th and the 20th centuries. In the southern hemisphere, where the first 'modern' catches were taken at South Georgia in the South Atlantic, humpbacks were the first species targeted: more than 22 000 were killed there in the first 10 years, from 1904 to 1913. Off Australia there was sporadic, but at times quite substantial, coastal humpback whaling off the north-west coast in 1912–1916 and 1922–1928, and on a larger scale, from pelagic whaling vessels, off the west coast in 1935–1939. From the east coast population there were catches from Jervis Bay, New South Wales, in 1912–1913 and a few at Twofold Bay, New South Wales, before 1930, but there was more or less continuous catching from New Zealand through to the 1960s. There was also pelagic whaling, south of 40°S, in the late 1920s and again in the late 1930s, with a substantial catch of over 2000 animals in 1940–1941. (See Tables 5.1 and 5.2).

Table 5.1 Humpback whale catches from Breeding Stocks D and E – north of 40°S

10-yr period	Breeding Stock D		Breeding Stock E			
	Australia NW coast	Australia SW coast	Australia E coast	Norfolk Island	New Zealand	Tonga
Pre 1914	1227		375		624	
1914–1923[1]	3973	1			823	
1924–1933	3435				864	
1934–1943	7223	20			742	
1944–1953	4024	271	1300	3	1059	
1954–1963	7255 (5)[2]	857	6123	884	1602	80
1964–1973	124 (124)					3
Post 1973						31
Totals	27261 (129)	1149	7798	887	5714	114
	28410 (129)		14513			

Table 5.2 Humpback whale catches from Breeding stocks D and E – pelagic, south of 40°S

10-yr period	Breeding Stock D	Breeding Stock E
Pre 1914		
1914–1923		
1924–1933	847	1125
1934–1943	5779	2080
1944–1953	3483	1969
1954–1963	11 517 (8469)[2]	28 043 (24 402)
1964–1973	396 (396)	509 (507)
Totals	22 022 (8865)	33 726 (24 909)

[1] 1923 = winter season 1923 and summer 1923/24, etc
[2] in brackets – previously unreported catches
Source: International Whaling Commission catch database

Australian humpback whaling began again after World War II, first off Western Australia at Point Cloates near North-West Cape (1949–1955), then at Carnarvon (1950–1963) and Albany (1952–1963). Catching from the east coast began a little later, from Tangalooma on Moreton Island, Queensland (1952–1962), and Byron Bay, New South Wales (1954–1962). Whaling on the same Breeding Stock (E) also took place at Norfolk Island (1956–1962) and in Cook Strait, New Zealand (1949–1962).

At the same time there was pelagic whaling on both populations in the Antarctic. The combined catches were so great that by 1963 the numbers were so reduced that it was uneconomical to continue Australian humpback whaling.

In his classic work on Australian humpbacks, Graham Chittleborough demonstrated that the west coast population (at that time known as 'Group IV' because its main feeding area was in Antarctic whaling management Area IV, 70°E–130°E) had been drastically reduced from an index of just under 0.5 humpbacks per 'catcher steaming hour' to just under 0.1, that is, by over 80 per cent, in 13 years. The same had occurred off the east coast (at that time known as 'Group V'), over an even shorter period, 10 years. By his calculations, the west coast population numbered some 12 000–17 000 animals prior to 1935. It had been reduced to about 10 000 by 1949, and to fewer than 800 by the end of 1962. His diagram showing the declines is reproduced here as Figure 5.4. He believed that to cause that huge reduction, particularly in the east coast population where there was a sudden decline of around 65 per cent in the three years 1959–1962, there must have been much larger catches than reported officially for those years – illegal, unreported, whaling must have been occurring.

Whales were indeed being killed illegally, by the Soviet fleets in the Antarctic, but on a scale even greater than Graham Chittleborough had supposed. The greatest numbers were taken from both stocks in the Antarctic in the years 1957–1961, but most from Breeding Stock E, particularly in the three years 1959–1961; over 12 000 were killed there in the single summer season 1959–1960. The catches, revised to include those unreported at the time, are summarised in Tables 5.1 and 5.2.

The most recent analyses (by the Scientific Committee of the International Whaling Commission) taking into account the most up-to-date information suggest that the population off the Australian west coast may have numbered about 20 000 originally, that is before about 1910, and that it had been reduced to about half that number by 1940. By around 1963 it probably numbered fewer than 1000 individuals. But from at least the mid-1970s, it was increasing, at a little over 10 per cent per year, to an estimated 10 000 in 1999. Recovery was at first delayed by the illegal, unreported, catches in the 1960s. The increase rate information comes from aerial surveys off Shark Bay, Western Australia, off Dirk Hartog, Bernier and Dorre islands, over the period 1982–1994 (see Figure 5.6), compared with aerial spotter data from the last commercial whaling season there in 1963. The 1999 population size comes from information from an aerial survey that year in the same area.

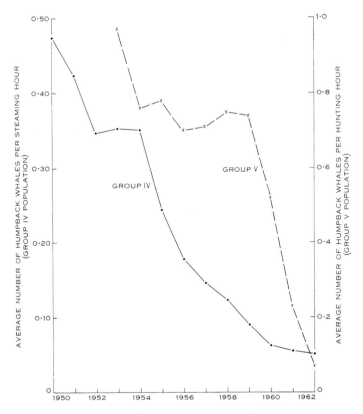

Figure 5.4. Relative abundance of humpback whales off the west and east coasts of Australia (Group IV = Breeding Stock D; Group V = Breeding Stock E, coastal substock). The plots demonstrate the enormous decline in abundance over the period, 1950–1962, but with a relatively greater decline over a shorter period (1959–1962) off the east coast, caused largely by illegal Antarctic catches.

Reproduced from Chittleborough, 1965, with permission from CSIRO Publishing

Off eastern Australia, the animals migrating along the east coast always seem to have been numerically fewer than off the west coast (Graham Chittleborough estimated them roughly in the proportion 1:1.5). Land-based surveys have been conducted since the mid 1980s from Stradbroke Island, Queensland, by two research groups, one at first under Professor Mike Bryden, University of Queensland and then under Dr Miranda Brown, the other by the late Dr Robert Paterson of the Queensland Museum and his wife Patricia, assisted by Dr Doug Cato of the Defence Science and Technology Organisation, Sydney, and continued since under Dr Michael Noad at the University of Queensland. Both have found a consistent increase over the period. The most recent estimate, by Michael Noad and

Figure 5.5 A humpback sighted between the Abrolhos Islands and Geraldton, 1977. Increasing numbers of humpback sightings off the coast in the 1970s led to a survey program to monitor subsequent population increase. The scar on the flank is assumed to be from illegal whaling operations in the 1960s. Photo: © Barry Wilson

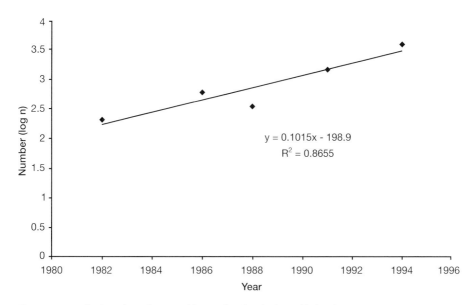

Figure 5.6 Relative abundance of humpback whales off Shark Bay, Western Australia, 1982–1994, showing the steady increase in the west coast population (Breeding Stock D) at about 10 per cent per year over the period. Reproduced from Bannister and Hedley, 2001, with permission from the Queensland Museum

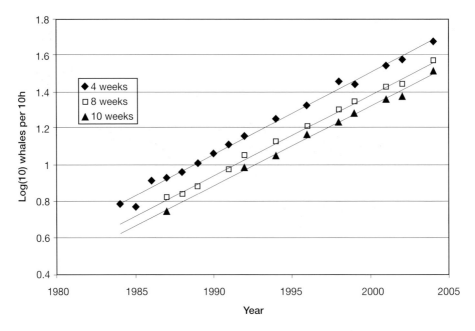

Figure 5.7 Increase (at around 10.6 per cent per year) in the average number of northbound humpback whales passing Point Lookout, Stradbroke Island, Queensland, per 10 hour period over the four, eight and 10 weeks around the peak of migration, each year 1984–2004. Courtesy of Dr Michael Noad, University of Queensland

his colleagues, shows an annual increase in that coastal migrating part of the population of around 10.6 per cent for the period 1987–2004 (see Figure 5.7), with a population estimate of some 7000 animals in 2004.

The assumption has been that coastal surveys, as off Shark Bay, Western Australia and Stradbroke Island, Queensland, covering the northern migration across the full width of the migrating stream, will accurately estimate total population size. However, an observed bias towards males in recent genetic studies on the breeding grounds and on migration has been taken as evidence that not all animals, particularly females, make the journey north each year. At present there is still argument about whether the observed sex-bias is real or the result of bias in sampling. Also, at least off Western Australia, recent evidence suggests that the survey may not entirely cover the migration stream.

Globally, humpback whales have been listed as 'Vulnerable' by IUCN, given their very reduced state as a result of whaling. Under the EPBC Act humpbacks are also listed as Vulnerable.

Southern right whale
Eubalaena australis

The southern right whale is robust and large – up to 17 metres in length. It has a large head, strongly arched upper jaws, long narrow baleen plates with fine inner 'fringes', callosities on the head, a distinctive 'neck', broad paddle-shaped flippers, triangular tail flukes with a smooth trailing edge, and no dorsal fin. Its blow is characteristically V-shaped when seen from in front or behind. Right whales were the 'right' ones to catch in the days of open boat whaling – they were slow-swimming and easy to approach, they floated when dead, and yielded massive quantities of product: oil for illumination and lubrication; baleen for corset stays, umbrella spokes, carriage hoops, even divining rods (see Chapter 3). Right whales are presumed to live for more than 50 years.

The body of a right whale is generally black although whales often have a white chin or belly splashes, and occasionally white or grey dorsal markings. Approximately three per cent of new born calves may be white, often with a black 'collar' and a pattern of black spots and dashes. The skin darkens to grey within the first year of life, and may then continue to darken, although the pattern of spots and dashes remains. Other grey or white marks on the back ('birthmarks') also persist throughout life. Head callosities (roughened warty patches of raised skin) occur at the tip of the upper jaw (the 'bonnet'), on the top of the head ('rostral islands'), along the upper edge of the lower jaw ('lip patches'), along the mandible and above the eye. Although initially greyish, they are home to large numbers of cyamid crustaceans ('whale lice') which give them a characteristic white or creamy colour.

Figure 5.8 A southern right whale 'blowing'. The V-shaped double blow is typical of the species. Photo: © John Bannister

The genus *Eubalaena* occurs worldwide, with three species now generally recognised: in the North Atlantic (*E. glacialis*), the North Pacific (*E. japonica*), and the southern hemisphere (*E. australis*). Their ranges do not overlap and evidence from genetics and parasites (particularly whale lice) – the latter studied by Professor Jon Seger at the University of Utah, USA – indicates very little if any current interchange between the species.

Right whales can be more active than obvious at first sight. Although cows with their young calves often seem to be idling close to the surf line in shallow bays, the calf is often quite active around its mother – both can even be seen breaching occasionally. Individuals can 'lobtail' (slap the surface with the tail flukes), and slap the surface with a flipper. One form of behaviour apparently unique to right whales, but of unknown function, is 'sailing' with the tail up and body perpendicular to the surface.

Studies of their behaviour by Dr Chris Clark of Cornell University, USA, have shown that right whales produce sounds related to social activity. As well as their lobtailing and flipper slapping which involves loud sounds, they also make 'blowing' sounds, as well as true vocalisations, which include 'up' calls seeking contact between individuals. As might be expected, animals in interacting groups are the most noisy; less active whales are much quieter.

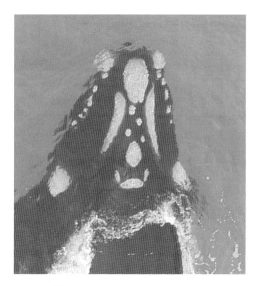

Figure 5.9 The pattern of callosities on the head is now widely used to identify individual southern right whales. This photograph taken from the air of an animal partly underwater shows a well-marked pattern that is very distinctive. Large catalogues now exist of animals photographed off the southern Australian coast. Photograph by Andrew Halsall © DEWHA

Distribution and movements

Southern right whales can be found in winter and spring in inshore waters of the major southern continents as well as off oceanic islands such as Tristan da Cunha in the South Atlantic, the Crozets and Kerguelen in the Indian Ocean, and Campbell and Auckland Islands south of New Zealand. They summer in colder waters, mainly down to around 45°S–55°S, but can be found further south. Off Australia they are usually seen between July and October mainly off the south coast between Cape Leeuwin in Western Australia and Ceduna in South Australia, and less commonly off Victoria and Tasmania. There are intermittent records on the east coast, even occasionally in Sydney Harbour. There are fairly frequent records of right whales on the west coast as far north as Shark Bay (around 26°S) with occasional sightings off Exmouth (around 21° 15′S), and on the east coast as far north as Hervey Bay (25°S).

Southern right whales have a habit of coming right in to the surf line in shallow water sometimes leading to concern that they are about to strand, but particularly for cows with their young calves that is their natural habitat at that time of year. If a large, dark, relatively inactive whale, with no dorsal fin, squarish flippers and a V-shaped blow, possibly accompanied

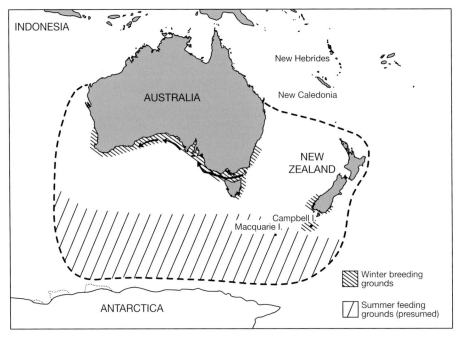

Figure 5.10 Presumed right whale distribution in the Australasian region. Also shown are some recorded movements along the southern Australian coast.
Modified from Bannister (1995)

by a calf, is reported 'logging' close to the surf, then it can't be anything but a right whale.

Right whales were familiar to early European settlers close to the Australian southern coasts in winter and spring, sometimes in large numbers. In the shallow parts of the Derwent River, Tasmania, 50 to 60 animals could reportedly be seen between May and November. They were such a hazard that it was said to be dangerous for small boats unless they kept close inshore, and there is even one report of the Lieutenant Governor being kept awake by 'the snorting of the whales at night'.

Evidence from the locations of 19th century pelagic catches, recent sightings surveys, Soviet catches in the 1960s and photographically identified individuals confirms right whale movements between the warm water breeding grounds and colder water feeding grounds. While most feeding, probably on copepods, may occur at around 45°S–55°S, along the Polar Front, there have been sightings of animals as far south as 65°S (and several 'matches' with animals recorded off the southern Australian coast); there were catches close to the same area by Soviet whalers in the early 1960s.

There is also evidence for anti-clockwise circular movement from cold waters towards the Australian coast in early winter. Observations by Dr Stephen Burnell, of the University of Sydney, at Head of Bight, South Australia, combined with information from aerial surveys along the southern Western Australian coast, have shown a predominately westward within-year coastal movement towards the end of winter and into spring before the animals move south in early summer towards the feeding grounds. There may be quite a lot of movement between localities in the same continent – for example between Argentina and Brazil, off South America, and in Australian waters between Tasmania and South Australia, as well as between South Australia and Western Australia. However there is generally very little interchange between land masses, although there are three recent records of movement between New Zealand and southern Australia.

Right whales, especially reproductive females, also show relatively high fidelity to a particular location on the coast. Not only do many of them stay in much the same area during the calving season, many may return to that area to calve in later years. By contrast, males or animals of unknown sex ('unaccompanied' animals) are much less predictable in their movements from year to year.

When moving from one location to another along the coast, right whales can be said to move steadily but surely, at not much more than three kilometres per hour. Stephen Burnell recorded one 'unaccompanied' animal moving approximately 350 km, between Nelson and Anglesea, Victoria, over less than four days at an average minimum speed of 3.7 km per hour.

Feeding and food

Based on the structure of their baleen, right whales have generally been regarded as feeding on smaller organisms than most other baleen whales. This is certainly true in the northern hemisphere where in the North Atlantic, for example, their main food is copepods. In the southern hemisphere most feeding, probably on copepods, may well occur at around 45°S–55°S, along the Polar Front, but in relatively cold waters, for example south of Australia (below 60°S) and around South Georgia in the South Atlantic, they must feed extensively on Antarctic krill (*Euphausia superba*). There are recent records of right whales feeding in much lower latitudes, for example off western South Africa, but off the Australian coast there seems to be little or no feeding, which in this part of the world must take place almost entirely in summer further south.

Right whales are 'skimmers', that is they swim through large swarms of their crustacean prey with open mouths. Water and prey enter through the front of the mouth and water is expelled out at the side, the prey being caught on the fine inner fringing hairs of the baleen. The highly arched upper jaw and the enormously expanded fleshy lips, combined with the very long narrow baleen plates, form a large filtering surface, quite different from the system employed in the rorquals where by contrast the mouth is capable of great expansion as they 'gulp in' large quantities of prey (see also Chapter 4).

Breeding
Mating in right whales is presumed to occur in winter, with gestation lasting around 12 months and birth in coastal waters mainly in July and August. Females seem to become sexually mature at around eight to nine years on average, and there is a well-marked three-year calving interval. Recent studies by Dr Russell Leaper and colleagues have shown that the breeding cycle in right whales wintering off Argentina may be lengthened to five years in poor feeding conditions, linked to changes in feeding ground (South Georgia) sea surface temperatures and El Niño. Weaning presumably takes place once the young are able to take solid food on the feeding grounds, although studies by Stephen Burnell at Head of Bight, South Australia, have shown that the cow may return briefly with the calf to its birth locality in the year after birth.

Cows accompanied by calves of the year are a special feature of several locations during winter and spring along the southern Australian coast. Cows may appear, mainly in late June or July – with a single calf being born soon after arrival – in places such as Doubtful Island Bay (especially Point Ann and Point Charles), and Israelite Bay, Western Australia, and at Head of Bight, South Australia, also to a lesser extent at Twilight Cove, Western Australia, near Victor Harbor, South Australia, off Warrnambool, Victoria, and on the south-eastern coast of Tasmania. They may then remain in much the same area for up to three months before leaving for higher latitudes in October or November. At the same time 'unaccompanied' animals, including juveniles and sub-adults of either sex, and adult males, may be found in interactive but unstable surface active groups ('SAGs') along the coast, often early in the season.

Successful breeding in right whales depends on sperm competition, with a number of males mating with a single female, each male producing a very large quantity of sperm from its enormous testes (in absolute terms the largest in the animal kingdom, at up to 900 kg the pair). In that respect, while the breeding regime is rather similar to that of humpbacks, it relies very much less than in humpbacks on direct aggression between males for a female.

Status

Southern right whale numbers were very severely depleted by open boat whaling in the 19th century and there was little sign of recovery in any populations until the mid 20th century. Until the 1950s, Australian right whales were regarded as virtually extinct. Graham Chittleborough's report of a sighting off Albany, Western Australia, in 1955 was the first he could find in the scientific literature that century. But increasing reports of their Australian occurrence, as elsewhere such as off Argentina and South Africa, suggested increasing numbers during the 1950s and early 1960s. Although illegal Soviet whaling in the 1960s temporarily prohibited recovery, most southern hemisphere populations have been growing strongly since then. Along the southern Australian coast between Cape Leeuwin, Western Australia, and Ceduna, South Australia (currently the headquarters of the Australian population), aerial surveys (at first off Western Australia 1976–1992 and extended into South Australia from 1993), have shown a recent rate of recovery, as elsewhere, of close to seven per cent, that is a doubling of the population every 10 years.

The total southern hemisphere population seems once to have numbered around 65 000. At least 150 000 were killed between the late 1700s and 1900, of which a remarkably large number, perhaps up to 60 000, were taken in the 10 year period 1830–1840. The population reached its lowest point in the 1920s, at around perhaps only 300 animals. Following recent recovery (interrupted by Soviet catches in the 1960s) the most recent southern hemisphere estimate is of around 7500 in 1997 (see Figure 5.11), but with the doubling time of 10 years it must be twice that amount now. Current Australian population size is estimated to be around 2000 animals along the coast between Cape Leeuwin, Western Australia, and Ceduna, South Australia, with a total Australian population of about 2500. That is still only a small proportion of its probable original population size. Dr Bill Dawbin, who worked on right and other whales, particularly humpbacks, from the 1950s until the 1980s estimated from studies of 'open boat' whaling logbooks that over 26 000 right whales were taken in south-western Pacific waters, including south-eastern Australia, between 1827 and 1899, with at least 75 per cent in the 10 years 1835–1844. Notably, the recent recovery has

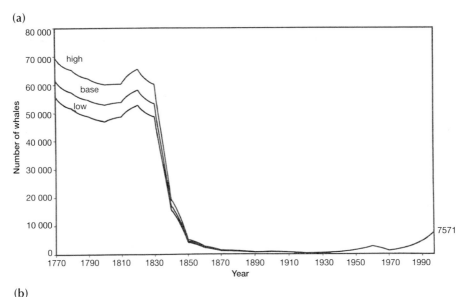
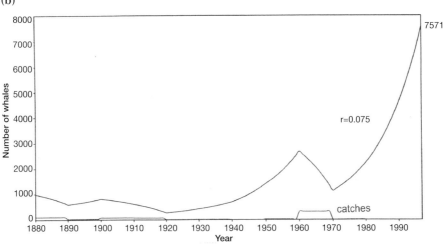

Figure 5.11 Population trajectory for southern hemisphere right whales from the early years of commercial exploitation, 1770–1997. The three scenarios, 'low', 'base', 'high', using three values for r, the population growth rate, indicate an original population at approximately 60 000, enormous and rapid reduction over the twenty-year period 1830–1850, and continued low levels until the mid 1930s, after which recovery began. Figure a) represents the total picture, 1770–1997; figure b) is the same, enlarged, for the period from 1880, but indicating catches to 1890, 1900–1920 and 1960–1970. The latter, taken illegally, caused a 10-year delay in the recovery seen from around 1960, resuming from 1970. At an increase rate of around seven per cent per year the southern hemisphere population should now, 10 years later, total around 15 000. Reproduced from International Whaling Commission 2001, with permission

not been in the area where those large catches were taken, but generally further west. Right whales can be seen nowadays in relatively large numbers at Campbell and Auckland Islands, south of New Zealand, but they are still rare around New Zealand itself, and, as already noted, generally uncommon off south-east Australia. It seems that whaling so reduced those in particular localities on the coast that they have not yet returned there, at least in anything approaching their original numbers. There is a nice example of this at the eastern end of the Great Australian Bight, where the southern Australian aerial survey records large numbers at Head of Bight, South Australia, each year (up to 50 or more calves recently), yet only 200 kilometres to the east, at Fowler Bay – a favoured 19th century whaling location – recent sightings have been intermittent, with at most only a few cows and calves each year.

The current estimate of Australian population size is based on data to 2006, but a disturbingly low number of calves was recorded in 2007 – at 57 cow/calf pairs it was the lowest for several years, compared with over 150 recently (there was a peak in 2005 of 177). Variation between years can be expected, particularly as each cow only comes on to the coast to give birth on average every three years. This results in 'cohorts' of different strengths, and although the 2007 cohort (last present in 2004) was not expected to be particularly high, it was not particularly low in 2004. What has caused the low Australian count can't be demonstrated clearly at the moment although El Niños in 2002–2003 or 2004–2005 may have been involved. The South Atlantic results reported by Russell Leaper and his colleagues were obtained from a long-term (30 year) sequence of counts; a similar long-term series may well be needed to provide similar conclusions for the Australian population. It will be important to try to obtain further annual counts in 2008 for comparison with those from 2007 and earlier.

Given their overall strong rate of increase off Argentina, South Africa and southern Australia, southern right whales have been listed by IUCN as 'Low Risk' (conservation dependent), although under the Australian EPBC Act they are classified as Endangered.

Sperm whale
Physeter macrocephalus

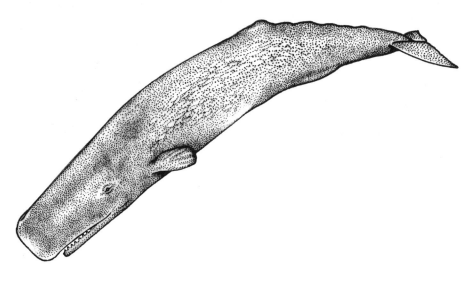

The largest toothed whale, the sperm whale is one of the most specialised whales, and among the easiest to recognise at sea.

The sperm whale has a huge boxlike head – in large males up to 40 per cent of the total body length. (The name *macrocephalus* means 'large head'.) On the left front of its head it has a single, asymmetrical blowhole. Its blow is low and bushy and projects diagonally forward. The sperm whale's body is dark greyish with an occasional ventral white splash, and white skin around the lips. A distinct whorl of white marks on the front of the head may enlarge in older males, together with other pale patches, to produce a much paler general colour, sometimes almost pure white as in the fictional 'Moby-Dick'. The dorsal fin is more of a low triangular 'hump' than a fin, being rather rounded and thick in cross-section. The relatively small flippers are paddle-shaped. The tail flukes are triangular and deeply notched; they are raised in the air ('fluked up') before deep diving. There are between 17–29, usually 20–26, erupted functional teeth on each side of the narrow underslung lower jaw, fitting into sockets in the upper jaw. Up to 11 small non-functional and often unerupted teeth can be found on either side in the upper jaw. The brain, averaging nearly eight kilograms in adult males, is the largest in absolute terms of any animal, although not particularly remarkable relative to the animal's overall size.

Figure 5.12 A school of sperm whales off Albany, Western Australia. The forward-directed blow, from the single blowhole on the left side of the head, is unique to sperm whales. Photo: © John Bell

Sperm whales are markedly dimorphic, the males reaching a maximum of 18 metres in length (though more commonly nowadays around 15 metres) and nearly 60 tonnes, while the females are up to 12 metres and about 25 tonnes. A thickened patch of skin, or callus, at the front of the dorsal hump has been used to identify adult females at sea. The name 'sperm whale' comes from 'spermaceti' (whale sperm) – the liquid wax filling the spermaceti organ ('case') in the head. Despite the size difference, both sexes can live to about 60 years.

Sperm whales strand frequently, sometimes in large groups. In Tasmania they are the second most frequently stranded species. A study of three sperm whale mass strandings during a single month, February 1998, by Dr Karen Evans of the University of Tasmania and colleagues reported that from 1911 to 2000 there were 69 Tasmanian sperm whale stranding events, of which 16 were mass strandings. Because of their size it is virtually impossible to rescue sperm whales once stranded on the beach. A rescue of seven adults out of twelve at Macquarie Harbour, Tasmania, in early March 2007 was successful because the animals were still in sufficiently deep water, between sand bars, to allow them to be netted and towed into deeper water.

Sperm whales make contact calls, social sounds, as well as generalised and identity codas as a form of communication. They also use high-frequency sound – trains of clicks punctuated by 'creaks' – to echolocate prey. They use the 'creaks' when homing in on their prey, much like bats in similar situations, but possibly not for their main food – deep sea squid (see also Chapter 4). Social codas, defined as patterned series of between three and 20 clicks lasting 0.5–2.5 seconds, are produced only in mixed social groups, not generally by adult males. Different family units develop distinct coda repertoires; the repertoire can also vary geographically.

The function of the head, more especially the spermaceti organ and its reservoir of oil, has been the subject of some conjecture. One view, the more traditional one, is that the spermaceti organ plays the same role in sperm whales as the 'melon' in dolphins, that is, to transmit and focus sound. But the extent to which the sperm whale actually uses echolocation for its main food, squid, has been questioned. While it may be that sperm whales echolocate fish, and not squid, they may use sound to orientate themselves in the water column both in relation to the surface and the sea bottom. The other view, propounded by Professor Malcolm Clarke of the Marine Biology Association, UK, is that the reservoir of spermaceti may act as a hydrostatic organ – as in a bathyscaphe – with water drawn through the blowhole cooling the oil and changing its density to allow rapid diving to great depths, and the heating of the oil through a copious blood supply then assisting a rapid ascent to the surface.

Distribution and movements
Found worldwide and generally in deep water, sperm whales approach land only where the coast is steep, such as near oceanic islands, or where the continental shelf is narrow, as off Albany, Western Australia; Kangaroo Island, South Australia; and off the Tasmanian west coast.

Females and juveniles are restricted to warm waters, warmer than about 18°C, between about 45°N and 45°S while adult males travel to and from colder waters in both hemispheres. Sperm whales have been recorded from all Australian states, but key Australian localities include a narrow area only a few miles wide at the edge of the continental shelf, about 20 to 30 miles offshore, between Cape Leeuwin and Esperance, Western Australia, and off Kangaroo Island, South Australia. They can be encountered off Tasmania, particularly on the west and north-west coasts; off New South Wales, including Wollongong; and off Stradbroke Island, Queensland.

Females and juveniles usually occur in 'nursery groups' of 20–25 animals, although larger aggregations have been documented. One report

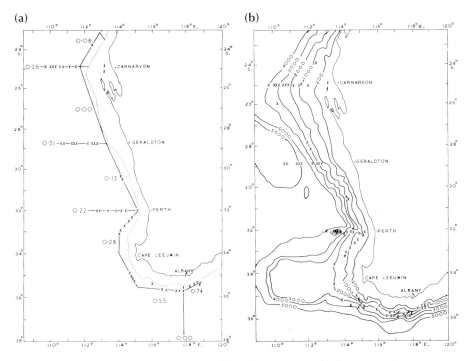

Figure 5.13 The positions of sperm whale sightings (x), and relative numbers seen, on the two-year aerial survey off Western Australia, 1963–1965. Legs were flown along the edge of the continental shelf (200 metre line) and out to 150 nautical miles off Carnarvon, Geraldton, Perth and Albany. Figure a) shows sightings widely distributed offshore north of Perth, but concentrated close to the shelf edge to the south, particularly off Albany. Figure b) relates that distribution to bottom contours: where the contours are far apart, the sightings were distributed largely offshore; where they are close together, as off Albany (and there is a rise from 5000–200 metres over only a short distance) sightings were concentrated close to the shelf edge. Reproduced from Bannister, 1968, with permission from CSIRO Publishing

from the Tasman Sea in February 1978 described animals 'spout[ing] from horizon to horizon' for over seven to eight hours, in a stream 70 miles wide. Males separate from the breeding schools in early adulthood, in 'bachelor' groups, and then travel to and from higher latitudes. The groups split up as the males grow older, so that in the Antarctic, only large males occur, moving as solitary or loosely aggregated individuals. The Albany whaling operation was based largely on 'bachelor males' found commonly with, but generally separate from, nursery schools. In the early 1960s, a two-year aerial survey in Western Australia along the edge of the continental shelf, from north of Carnarvon to east of Albany (see Figure 5.13), showed sperm whales distributed well off the coast where the shelf was wide (up to almost 200 km) on the west coast north of Perth. Off the south coast where the shelf itself was narrow (about 40 km wide) they were concentrated in only a narrow strip along the continental slope. Movement through the whaling area, south and east of Albany, was predominantly from the east – 84 per cent of over 800 observations were of animals moving to the west, following the shelf line. Seawards (to the south) of that narrow area sperm whales were hardly present.

Feeding and food

Sperm whales generally feed on medium-sized (0.20–1.00 metre) oceanic squid, although females eat smaller individuals than do the larger males. From items found in the stomach, such as sponges, rocks and sand, much feeding must take place on or near the sea floor. Adult males can dive to great depths, certainly to over 1000 metres, remaining down for an hour or more. Before such dives they usually raise their tail flukes ('fluke up'); during them, they may descend and ascend almost vertically. After such dives, they remain at the surface, blowing repeatedly, behaviour known by old-time whalers as 'having their spoutings out' – the number of 'blows' was said to be equivalent to the number of minutes spent under water.

At the Albany whaling station, over 70 per cent of the animals examined had been feeding recently, almost wholly on squid. Less than two per cent had been feeding on fish, mainly deep sea forms such as angler fish. Remains of giant squid (*Architeuthis* sp.) longer than two metres were found occasionally. The heavy feeding recorded off Albany was correlated with the very steep continental slope there (depths going from 200 metres to 5000 metres over only few kilometres) coupled with the presence of submarine canyons. These features are also present, for example, along the shelf south and south-east of Kangaroo Island, South Australia.

Breeding

Sperm whales are highly social animals. The basic element, the family unit, comprises a group of some 10 to 15 adult females and their young. Large males roam between the nursery groups, seeking receptive females. The males are usually solitary but, as already noted (see Chapter 4) judging from tooth scars on the head must occasionally fight other males, presumably for possession of the females.

Unlike most other great whales, sperm whales have a low birth rate and are relatively 'late developers'. From data obtained at the Albany whaling station, the main pairing season is in early summer (September to December) with gestation lasting just under 16 months, and calving mainly from mid-summer to autumn. Females become sexually mature on average around 12 years old, at a length of about nine metres. The calving interval at Albany was estimated as three years; elsewhere it has been recorded as between four and six years. Calves may take solid food after about a year, but some suckling by the mothers may extend for several years. Males on the other hand, while they may become sexually mature at around the same length and age as females, that is, at about 11 years and eight to nine metres, do not then take an active part in breeding. They only become 'socially mature' much later, from around 25 years old.

Status

Sperm whales were taken commercially worldwide from the 1700s. Off Australia most such 'open boat' whaling was by American vessels in the 19th century, with activity peaking between 1830 and 1850. Major Australian hunting areas were the 'Coast of New Holland' Ground off Western Australia and the 'Middle' Ground in the Tasman Sea. On the Coast of New Holland Ground sperm whales could be taken all year round while whaling on the Middle Ground was mainly in summer (December to March). The only major 20th century Australian 'modern' sperm whaling operation was off Albany, Western Australia, from 1955 to 1978 (see Chapter 3). There had been earlier sperm whaling there, at Frenchman Bay, between 1912 and 1916, largely in summer. The post-World War II Albany operations, however, extended over the colder months, from February/March to November/December, but their timing was initially influenced by the local availability of humpbacks during winter. Concerns over excessive reduction of breeding males and a possible pregnancy rate decline contributed to Albany's closure in 1978 (see Chapter 6). Although subject to national quotas based on assessments by the International Whaling Commission's Scientific Committee, the 'Albany stock' was assessed as having undergone a major decline over the period 1947–1979, with males (over 20 years old) depleted by 91 per cent and females (over 13 years old) by 26 per cent over the period. For assessment purposes sperm whales in the south-east Indian Ocean, including those off Albany, were considered to belong to a single management stock ('Division 5' of nine southern hemisphere divisions), with a northern limit north of 20°S, and longitudinal limits 95°E in the west to Tasmania (around 147°E) in the east, with females and young males restricted to north of 40°S–50°S and adult males travelling to and from the

Antarctic. Some confirmation of the stock caught off Albany being related to animals further east came from the disruption of catching off Albany by Soviet pelagic operations further east in March 1965.

IUCN has listed sperm whales as Vulnerable.

6
CONSERVATION AND REGULATION

To account for the steady increase of [whaling industry] production … we must look not to an improved technology or any resulting increase in output … but to a multiplication of vessels employed in that production … As vessels increased in number, the stocks of whales on the most accessible grounds declined. This was, of course, an ancient pattern, and in the 19th century, as in earlier times, the history of the industry can be seen as a history of the search for new and more productive whaling grounds, where the familiar types of whales could be exploited. The successive opening of these grounds offered the whalemen sufficient opportunity to meet the needs of the market …
  Dr Richard Kugler, Director of the New Bedford Whaling Museum, USA, 1971.

Whales migrate far beyond national boundaries and therefore effective management – which may range from full protection to controlled whaling – requires international cooperation. This is complicated by

the seasonal concentration of many whales in the Antarctic ... Some nations regard whales as a renewable resource which they are free to exploit; others are opposed to the killing of whales. So an internationally accepted policy, prudent enough to meet the major concerns of nations opposed to whaling, is needed for the regulation of whaling.
Sir Sydney Frost, Report of the Independent Inquiry into Whales and Whaling, 1978.

There is much work to be done on the stocks ... and there is a wide field of general research on whales as one of the resources of the sea, but the immediate difficulties of regulation are now attributable to political and economic factors and not to lack of scientific evidence.
Dr N. A. Mackintosh, The Stocks of Whales, 1965.

Traditional, open boat, whaling was not subject to any controls other than the availability of whales, market forces, and the vagaries of wind and weather at sea. It was literally every whaleman (or rather captain) for himself, and as quoted from Dr Kugler above, the result was a continuing search for new grounds to discover and new populations to exploit. The same was true for the first years of modern whaling. Indeed the first attempt at any regulation (other than licensing of shore-based operations, circumvented by the development of the pelagic factory ship) was not until 1931, 27 years after southern whaling had begun at South Georgia. Although a measure was agreed for protecting right whales (actually implemented in 1935) it was the fear of low oil prices arising from over-production that led to moves to limit catching, with whaling vessels to be licensed and catch statistics to be collected. Although largely ineffective the measures did establish the principle of international regulation. Next came international conferences in 1937 and 1938 where there were major advances: in setting minimum lengths for each species, totally protecting right (and gray) whales, temporarily protecting Antarctic humpbacks, limiting factory ship operations to waters south of 40°S (and part of the North Pacific), establishing an Antarctic sanctuary between 70°W and 160°W (south of the Pacific), determining opening and closing dates for an Antarctic season, and arranging for inspectors to be appointed to each factory. An attempt was also made to limit the total catch, but that had to wait till after World War II. Most unfortunately however, when that limit was adopted, by the newly founded International Whaling Commission (IWC) in Washington in 1946 (and brought into effect in 1948), it was based on the fatally flawed Blue Whale Unit (bwu). One bwu was equivalent to one blue whale, two fin whales, two and half humpbacks and six sei – based on average oil yields. But an overall catch limit in bwu made no allowance for the particular status of any species or population – once most blue whales had been taken, attention could turn to fin whales, then sei, and/or humpbacks, or any combination of the species, without regard to their individual conservation status. Essentially, one or more species or stock could be rendered 'commercially extinct' while operations continued on other species or stocks within the overall bwu system.

So despite the overall limits, stocks at first of blue whales, then fin – humpback numbers had already been considerably reduced, especially in the South Atlantic sector – and later sei, were successively depleted. At the same time, there was no independent enforcement or inspection system (inspectors were appointed by the country whose operation they were to inspect), and the whaling nations were unwilling to recognise the increasing evidence of deterioration in the stocks, despite the evidence being accumulated by the Commission's scientific advisors, meeting each year as its Scientific Committee. Things came to a head in the early 1960s when the Commission appointed an independent 'Special Committee of Three' (later four) scientists, from nations not engaged in Antarctic whaling, to assess the Antarctic stocks. Their results, available in 1963, showed clearly that blue whales and humpbacks needed total protection, and that the catch of fin whales needed to be severely curtailed. The whaling nations were unable to accept the latter, however, and no quota at all was set in 1964; the situation then remained extremely unsatisfactory for several years, indeed until 1972, when an international observer scheme was at last implemented. But by that time fin whale stocks were severely depleted, and attention had already turned to the much smaller sei whale. But here it is time to consider the Australian situation.

Whaling on 'Australian' humpbacks has already been described in Chapters 4 and 5. Humpbacks were given full protection in the Antarctic for the period 1939–1949 (but with an exception in 1940–1941). Catch restrictions were brought in for the Australian operations that began after World War II. From 1949 on the west coast and 1952 on the east coast separate annual catch quotas were set by the Australian Government for each whaling station. The quotas varied from 500 to 600 at Point Cloates until 1955, when its quota was transferred to Carnarvon, itself allocated a total of 1000 until 1960 and then lowered progressively to 400 in the last year of whaling there, 1963. At Albany the quota varied between 50 and 175; in its last year of humpback whaling, also 1963, the quota was 100. Similar numbers were allocated for east coast operations – 600–700 at Tangalooma, Queensland, 1952–1962; 120 at Byron Bay, NSW, 1954–1962; and 120–170 at Norfolk Island, 1956–1962. In the Antarctic a limit of 1250 humpbacks for the whole area was set for each summer season from 1949–1950 to 1951–1952, except in the Pacific sector where the area 70°W–160°W was reserved as a sanctuary. From 1952–1953 the catch limit was set not as a number but by limiting the catch to a four-day season, and from 1954–1955 the sanctuary was enlarged to include the area between 0° and 70°W (but with a minor variation to alter the western boundary to 60°W from 1958–1959). As already noted, humpback whaling was prohibited altogether from 1963. But also as already noted, there was considerable illegal, unreported, whaling, especially on humpbacks but also on other species, until the advent of the International Observer Scheme in 1972.

The Albany, Western Australia, whaling station, starting with humpbacks in 1952, had been taking them as well as sperm whales from 1955 to 1963, but it caught only sperm whales from 1964 – by then it was the sole remaining Australian whaling operation – until its closure in 1978. The International Whaling Commission set overall sperm whale catch limits for the southern hemisphere from 1971; they were set separately by sex from 1972 and by area from 1973, and mirrored domestically by the Australian Government. Sperm whales taken off Albany came within the IWC's designated southern hemisphere Division 5 (90°E–130°E). For 1973, Albany's quota was 900 males and 505 females, of which it actually caught 684 and 287. Over the ensuing period, the quotas were reduced; in 1978 the catch was 509 males and 170 females. There were also changes in the size limits, resulting from the application of more sophisticated assessment models (see below), to 30 feet[2] from a minimum at land stations prior to 1972 of 35 feet, and with a maximum limit of 45 feet applied later for males. The latter was for the area between 40°S and 40°N during the breeding season, thus affecting Albany, situated at 35°S.

Australia had been an original signatory to the International Convention on the Regulation of Whaling developed at the 1946 Washington Conference, which established the IWC – responsible for the conservation of whale

---

[2] Despite the use of metric units by many IWC member nations, the measurement unit has traditionally remained as the standard English foot.

stocks and the management of whaling. The Commission had, as has been seen, regrettably little overall success, undertaking too little too late, or, as Greg Donovan, currently Head of Science at the Commission, has put it – 'IWC actions, while insufficient, were better than nothing … [but at least from] the 1970s its actions were based largely on scientific advice, to a degree probably unparalleled in any fisheries commission'. But a radical change in IWC activities took place in the early 1980s. From the mid-1970s its composition had been changing, initially from only whaling nations, to a mix including non-whaling nations whose main interest was whale protection. By 1982 the mix had so changed that in that year the necessary majority adopted a moratorium on all commercial whaling, to take effect from 1986, initially for 10 years, but subsequently maintained. In that period Australia's attitude also had altered radically, indeed by 180 degrees, as a result of a change in government policy based on the findings of an independent judicial inquiry under The Hon. Sir Sydney Frost, appointed in March 1978.

The Frost Inquiry (as it came to be known) resulted from mounting Australia-wide public pressure and concern in the context of a vociferous worldwide environmental movement in which whales were seen as special creatures that shouldn't be killed; where whaling was endangering some species to the extent that there were fears some might become extinct; where hunting methods were seen to be primitive if not barbarous, and substitutes were mostly, if not wholly, available for whale products. The Inquiry's main task, in the context of Australia wanting to preserve and conserve whales (including cetaceans generally), was essentially to decide whether Australian whaling should continue or cease. Its report (the Frost Report), presented in December 1978, had as its main, and plain, conclusion that Australian whaling should end, and that internationally Australia should pursue a policy of opposition to whaling.[3] In doing so however Australia should remain a member of IWC: it should seek a worldwide ban on whaling; it should continue to support Australian research on cetaceans; and it should continue to be involved in the Commission's Scientific Committee. The Government adopted all the report's conclusions and recommendations, and replaced the *Whaling Act 1960*, with the *Whale Protection Act 1980*. Thus from an active whaling nation Australia became, almost overnight, an active non-whaling nation. The provisos for Australia's

---

3   The Frost Inquiry concluded that the only whale products used in Australia that could not be readily replaced were sperm oil and spermaceti, but that suitable substitutes could become available within a period of two years.

continuing IWC involvement were put into effect and have persisted. Australia has continued to be active on the Scientific Committee: among other things it has provided three committee chairs (one of them twice) and several convenors of subcommittees. The government was also to the forefront in promoting the formation of the Southern Ocean Sanctuary for the specific protection of cetaceans from commercial whaling in those international waters. Under the 1980 Act cetaceans were protected within Australian territorial waters (the 200 nautical mile Exclusive Economic Zone), and Australian citizens, worldwide, were bound by its provisions. More recently, that Act has been superseded by the comprehensive *Environment Protection and Biodiversity Conservation Act 1999* (EPBC Act), covering all wildlife species, and with special provisions for cetaceans. The provisions include strong protection for all species; identification of key threats; protection of critical habitat; preparation of recovery plans; and regulation of export and import of live animals, specimens and products. Of special interest in the present context, recovery plans have been prepared so far for blue, fin, sei, humpback and southern right whales. The Act continues to apply to Australian citizens, worldwide.

Returning to the wider international scene – Graham Chittleborough's concerns over possibly misreported humpback catches have already been mentioned (see Chapter 5, Humpback whale), together with the revelation that the illegal catching had been on a larger scale than even he had imagined. In fact – as could not be revealed until after the end of the Cold War, but was publicly reported at the biennial Meeting of the (international) Marine Mammal Society in November 1993 – the Soviet Union had been conducting a secret campaign of illegal hunting from the late 1950s, taking virtually everything it came across, regardless of species, sex, size or protected status. The official statistics, submitted to the Commission, had been doctored to hide the real situation. But from the actual data obtained on board the factory ships by individual scientists, and hidden by them at considerable personal risk for more than 20 years, it was shown that in the southern hemisphere (similarly in the north but on a smaller scale) over 100 000 more whales had been killed than notified officially. Although humpbacks were the main victims, other species were affected, such as sperm whales (see Chapter 5, Sperm whale, for the effect of such catches off Albany in 1965); large catches of pygmy blues off southern Australia and into the Indian Ocean went unreported, as well as southern right whales (for which a delay in recovery in the 1960s – see Chapter 5, Southern right whale and Figure 5.11 – has been attributed directly to illegal Soviet catches in those years). The illegal catching only ceased with the introduction of the IWC's International Observer Scheme in 1972, by which time the stocks of the larger species had anyway been virtually exhausted.

While catching continued on dwindling stocks in the 1970s considerable concerns were expressed over the rationale for scientific management. Worldwide there was growing concern over whaling generally and this received a major boost at the 1972 United Nations Conference on the Human Environment (the Stockholm Conference) and led towards the end of the decade to the Australian (Frost) judicial inquiry (see above). The result was the adoption in 1975 of the Commission's New Management Procedure (NMP), based essentially on determining the maximum sustainable yield (MSY) for a particular management stock; one then only had to know where the stock size was in relation to that level to be able to determine whether whaling could continue, either on a stock not previously harvested (classified as an Initial Management Stock, IMS), or at a continuing level (as a Sustained Management Stock, SMS), or should be altogether prohibited (as a Protection Stock, PS). The system was excellent in theory. A major element was that it removed political considerations from the Commission's catch level decisions – once the Scientific Committee had the required information, the required classifications followed automatically. But in practice it proved very difficult to apply. In many cases information was just not sufficient to determine stock boundaries, or, when it was (or they could at least be reasonably estimated), there was often not sufficiently robust information on current numbers. Also there was a virtually impossible load of stocks requiring assessment. Into the early 1980s the Scientific Committee struggled to apply the procedure to various stocks, for example southern sei whales, and also sperm (for which a specific and exceedingly complex assessment model was developed), but in 1986 it was overtaken by the moratorium under which all commercial whaling was prohibited.

In the meantime the IWC's Scientific Committee continued to explore development of a more practical management procedure, resulting in 1994 in the Commission's adoption of the Revised Management Procedure (RMP). This depends essentially on a knowledge, once more, of stock boundaries, also of current abundance (from sightings surveys), and of past and present 'non-natural' removals, including any catches as well as by-catch in fishing gear and deaths from ship-strikes. With intensive computer trials the risk of various scenarios can be assessed, and using a Catch Limit Algorithm (CLA) the most conservative catch limits can be determined. The CLA involves 'feedback' from accumulating information from sightings surveys (and any catches or other removals). Again as Greg Donovan has put it 'The RMP sets a standard for the management of all marine and other living resources'. But for its implementation it has to be incorporated into a Revised Management Scheme (RMS), involving such non-scientific concerns as inspection and enforcement. So far the Commission, with its current composition of whaling and non-whaling members (see below) has been unable to reach agreement on its adoption. Meanwhile, Australia, in line with its non-whaling policy, has remained firmly opposed to any return to commercial whaling.

Between 2000 and 2006, the Commission saw something of a shift towards a fifty-fifty mix of whaling and non-whaling interests, with an enlarged membership. In 2000, approximately one-third of the membership supported whaling, while at its 2006 meeting, a resolution seeking a ban on the moratorium was supported by 33 members, opposed by 32, and there was one abstention. But in 2007, with an even more enlarged membership, the mix had swung back somewhat towards 'non-whaling' – 39 nations (55% of those present) supported a proposal to establish a whale sanctuary in the South Atlantic. Among several contentious issues – including attempts to establish a whale sanctuary in the South Pacific – one has attracted much public attention: 'scientific permit' or 'research' whaling. This has been of special concern to Australia, since those catches have been concentrated in Antarctic waters south of the continent, and even more especially because of proposals for including, from summer 2007–2008, catches of humpback whales. Not unnaturally Australians regard these as 'their' whales, and therefore in special need of protection, given that they migrate along either side of the continent and breed in its northern waters.

However much people may deplore it, scientific permit whaling is allowed under the Convention, where member governments may issue permits for the killing of whales for scientific purposes. Prior to the moratorium such permits had been issued on a number of occasions, including by Australia for the catch of 'small' sperm whales (and therefore to include females, as well as young males, otherwise mostly excluded from the commercial catch). But the question has been particularly contentious following the moratorium, where Japan, Norway and Iceland have issued such permits, and the accusation has been made, particularly against Japan, that they provide a means of circumventing the moratorium and are of little scientific justification. The situation has been exacerbated by the fact that the products, particularly meat for human consumption, are sold on the open market, even though there is a provision in the Convention that the animals should be utilised once the scientific data have been obtained. In Japan's case, (under a project known as the Japanese Research Programme in the Antarctic, JARPA) some 400 Antarctic minke whales were killed under permit each year from the summer season 1987–1988, and alternately in Antarctic Areas IV and V, that is, south of Australia between 70°E-130°E and 130°E-170°W respectively. Under a revised proposal, 'JARPA II', taking effect from 2005–2006, the numbers were increased to a maximum per year of 850 minke, plus or minus an allowance of 10 per cent, and 10 fin whales, for a two-year feasibility study (the catch in 2005–2006 was actually 853 minke and 10 fin). From 2007–2008, when the program is to come fully into force, catches were to be extended to include 50 humpbacks and 50 fin whales (but see Chapter 7). Initially, JARPA was justified on the grounds that it would provide information important in stock assessment, such as natural mortality rates obtained from the whale's individual ages, only obtainable from earplugs (see Chapter 4), extracted from dead animals. More recently, emphasis has turned towards ecological research, and especially in JARPA II, including, controversially, food and feeding – one reason given for including humpbacks and fin whales, whose increasing numbers are considered by the proponents as possible competitors with minke whales. While the Commission is not required to approve such permits, they are subject to Scientific Committee review when issued, and to periodic review at intervals. The JARPA minke whale program has now been reviewed twice, most recently in 2006. On each occasion the review has stated that while the results are not in fact *required* (author's italics) for management of minke whales under the RMP, they do have the potential to improve the management of minke whales (should they eventually ever be taken commercially) in a number of ways; they could even increase the catch without increasing the risk of depletion of the stock above a level of risk determined for that stock under the RMP.

7
THE FUTURE

The range of threats facing marine mammals is vast, as is our ignorance about how to evaluate and manage those threats. On one hand, there are situations [where] the needed actions are fairly obvious: close a fishery; stop a hunt; eliminate motor vessel traffic in areas where the animals rest, socialise, feed or nurse their young. More often than not, the political, economic and cultural dimensions are more complex and contentious than the biological or ecological ones. On the other hand, there are situations [where] the conservation imperatives are obscured by genuine scientific uncertainty … Why is a whale population not reproducing at the expected rate …? A necessary step … is to understand the cause or causes of population decline.

R R Reeves and P J H Reijnders, Conservation and Management, in Marine Mammal Biology: An Evolutionary Approach, A R Hoelzel (ed.), 2002.

Any discussion of what the future may hold for Australia's great whales has to include a review of current and future threats, of what might be done to mitigate them, and of the research needed to provide the information on which any protection or conservation measures must be based. For cetaceans in Australian waters these topics are currently addressed under the operations of the *Environment Protection and Biodiversity Conservation Act 1999* (EPBC Act).

The EPBC Act provides for the formulation of recovery plans for individual threatened species. Each plan aims to maximise the long-term survival of such species in the wild, by detailing both what has to be done to protect and restore important populations as well as to manage and reduce threats. So far plans have been prepared jointly for blue, fin and sei whales, and separately for humpbacks and southern right whales. Each plan outlines its objectives, criteria to measure its performance against the objectives, information on each species (referring back to a more detailed Species Profile, available on the web and regularly updated), description of critical habitat, and details of domestic and international management. It also details threats – both current and potential (referring back also to a Threats Database available on the web) – and actions to achieve the objectives. The latter cover programs to measure population recovery, to define critical habitat, and to provide protection from threats. Also detailed are any major benefits to other native species or communities, the duration and costs of each program, the role and interests of indigenous people, and any social and economic impacts of the plan. Finally there is a list of organisations and/or persons involved in evaluating the plan's performance.

Each recovery plan so far prepared includes the twin objectives of a) recovery to the point where the population can be considered secure in the wild (defined as having 'sufficient geographic range and distribution, abundance and genetic diversity to provide a stable population over long time scales') and b) maintenance of protection of the species from anthropogenic (human) threats. In the case of humpbacks and southern right whales, both (in contrast to blue, fin and sei, as already seen) the subject of considerable over-exploitation in Australian waters, there is an additional objective, aimed at ensuring eventual distribution of the species similar to that before exploitation.

Each plan's effectiveness is to be judged according to the species involved. For the two populations currently recovering (humpbacks and southern right), there should be evidence that they are continuing to do so at the optimum biological rate – as has appeared to be the case. For blue, fin and sei, the requirement is to obtain an indication of current population size and recovery status for animals using Australian waters. All species should have a distribution similar to, or expanding towards, their original state. At the same time domestic and international protection should be maintained or improved.

Under identified threats, two are currently recognised: a) the resumption of commercial whaling and/or expansion of scientific permit whaling and b) concerns over habitat degradation. For a), proposals under JARPA II (see Chapter 6) which include 50 humpbacks in the Japanese special catch in the Antarctic south of Australia from 2007 have caused considerable public concern.[4] A take at that level will have little or no effect on the west coast population (Breeding Stock D, see Chapter 5, Humpback whale) as such, but where mixing of populations or subpopulations on feeding grounds may be occurring, as south of eastern Australia (that is, from Breeding Stocks E or F, see Chapter 5, Humpback whale), any removals could adversely affect those relatively small stocks, particularly where no recent increase has been detected. Under b), threats identified include acoustic pollution (vessel noise and seismic activity), entanglement, ship-strike, infrastructure development (marinas, wharves, aquaculture, mining/drilling), as well as changes in water quality and pollution. Given their use of Australian coastal waters, blue whales might be threatened by any or all of these, but the greatest concern has to be for humpbacks and right whales, given both species' dependence on Australian inshore areas for migration and breeding. The plans recognise that habitat degradation could result in reduced occupancy and/or exclusion of individual animals from suitable habitat, as well as possible reductions in reproductive success, and even mortality. But both species are currently increasing at or close to the optimum biological rate, so habitat degradation does not seem to have had a negative impact on the populations as a whole, although there are likely to be lag effects. Nevertheless, two activities are currently of particular concern – seismic activity and entanglement.

Given the considerable, and increasing, exploration activity around the Australian coast – for example at present off the north-west coast and in Bass Strait – detailed guidelines have been developed under the EPBC Act to address interactions between offshore seismic operations and certain listed cetaceans, including all the Australian great whales. Such operations are regarded as likely to have a significant impact where the operation is undertaken within 20 kilometres of a cetacean feeding, breeding or resting area, or in some circumstances in or near their migratory paths. A permit is required where a

---

4   In late 2007, in the face of considerable public pressure led by the Australian Government, Japan decided not to implement its plan to kill 50 humpbacks in 2007–2008.

seismic operation will interfere with a cetacean. Management guidelines have been developed to minimise interference, detailing procedures to be followed during the operations. In particular, there must be observations for the presence of whales, beginning at least 90 minutes before any high-energy acoustics are employed, focusing on an area within three kilometres of the survey vessel. An obvious way to avoid interference with southern right whales and humpbacks in Australian waters is of course to ensure where possible that activities are planned for outside the migration/breeding season (essentially winter and early spring).

Humpbacks and to a lesser extent right whales are particularly susceptible to entanglement, in shark nets, pot lines, or aquaculture installations. Disentanglement measures have been developed and are reviewed at annual workshops attended by relevant State and Federal authorities. For 2006, 16 entanglements were reported, from Queensland, Victoria, Tasmania and Western Australia, involving 14 humpbacks, one southern right and one pygmy right whale: eight animals were successfully disentangled; there was one death and the fate of the remainder is unknown. While such impacts obviously affect individual animals, and not, at that level, the population as a whole, should they occur more intensively or over a larger portion of the animals' range they could lead to broader impacts, for example by reducing recruitment and impeding population recovery.

In addition, special concern has been expressed in the case of right whales where the relatively few that visit the south-east coast (less than 10 per cent of the total Australian population at present) may be subject to the larger and increasing human population pressures there.

Elsewhere, for example in the North Atlantic and the Mediterranean, fin whales have been reported as subject to ship-strikes. As southern populations increase collisions with ships could be expected to increase in major shipping lanes off the Australian coast.

As for potential threats, two are of greatest concern: climate change and prey depletion through over-harvesting. Climate change is likely to result in reduced Southern Ocean productivity and oceanographic changes, the former leading to fluctuations in food availability and the latter to altered habitat. Reproductive success in sperm whales near the Galapagos Islands in the eastern central Pacific has been linked to warmer sea surface temperatures. In the North Atlantic atmospheric and oceanographic variations are influencing right whale calving through effects on the food supply; continuing global warming is likely to have further impacts on those resources. In southern right whales a correlation between Southern Ocean sea surface temperature anomalies and reproductive success has recently been demonstrated (see Chapter 5, Southern right whale), suggesting that as Antarctic feeding grounds warm up the average calving rate of southern right whales may be expected to decline. The low number of calving right whales recorded off southern Australia in 2007 may have resulted from similar sea surface temperature changes (again see Chapter 5, Southern right whale). But as Dr Robert Kenney of the University of Rhode Island, USA, has suggested – for North Atlantic right whales, but his comments are applicable more broadly – they 'have survived for a very long time, through ice ages, warm periods and everything in between. Because they live a long time it may be that [they] can wait out the bad periods and thrive in the good ones.' The problem is predicting the speed of any changes that do occur. In the meantime, again in Dr Kenney's words 'there is hope that the resilience … that enabled … right whales to survive in the past [including the depredations of whalers] will also serve them well in the face of a warming and changing ocean.'

If prey depletion through over-harvesting were to become a major factor, there would certainly be effects on blue, fin, humpback and Antarctic minke whales, all of which are almost entirely dependent on Antarctic krill as their staple diet, but lesser effects on sei and right whales could be expected. Currently however, krill harvesting is managed on an ecosystem basis under the Convention for Conservation of Antarctic Living Resources (CCAMLR), taking predators' (and thus whales') needs into account, and removals are at present at a low level. Concerns over how ecological and climate-related processes interact to affect ecosystems have led to development, largely within CCAMLR, of a Southern Ocean Integrating Climate and Ecosytems Dynamics (ICED) program, greatly dependent on long-term (more than 25 year) and large scale (for example, circumpolar) studies. The International Whaling Commission will be holding a workshop on climate change and its impact on cetaceans after its 2008 annual meeting.

For those great whales not so far the subject of recovery plans, the most likely threats include entanglement and ship-strikes, especially for Bryde's whales and possibly dwarf minkes. Bryde's whales might be affected by overfishing of prey species such as anchovy. There are also occasional reports of sperm whales colliding with ships, and there is a possibility of sperm whale entanglement in deep sea gillnets.

To achieve the objectives, a number of actions are proposed under the recovery plans for each species. They include monitoring population recovery, defining characteristics of critical habitat, ensuring protection from threats – including from directed take (commercially or under scientific permit) – directly protecting critical habitat, monitoring and managing the potential effects of prey harvesting, and monitoring climate change. It is also recognised that given the status of the various populations, each recovery plan will take longer than the five years formally allotted to it.

Successfully undertaking the necessary actions requires research at various levels. For Australia's great whales research has been, or is being, attempted in a variety of ways. For each species, a question addressed under the Australian Cetacean Action Plan (published in 1996) was 'Can research … be carried out with existing resources?' Not unexpectedly, for no great whale species was the answer an unequivocal 'yes'. Particularly for the mainly oceanic species – blue, fin, sei, Antarctic minke – it was recognised that considerable reliance would have to be placed on cooperation with overseas agencies, such as the IWC undertaking Southern Ocean sightings surveys, or CCAMLR conducting ecological studies. For the coastal species – humpbacks and southern right whales – Federal Government funding had been available for ongoing monitoring (in the case of humpbacks off Queensland and Western Australia, and for right whales off southern Australia including Western Australia and at Head of Bight, South Australia), while several short-term and specific projects, including genetic studies for assessing relationships with other populations, had been funded from non-government sources. For each species there was a particular need for dedicated funding over specified, long-term time frames to undertake population assessment and monitoring, as well as photo-identification studies. For sperm whales, there was a proposal to replicate, as far as possible and at appropriate intervals, the commercial aerial spotter operations off Albany, Western Australia.

Up to the present, monitoring of humpback and southern right whales has generally continued, but mainly on a year-to-year basis, funded from the Natural Heritage Trust through the Commonwealth Department of the Environment and Water Resources (now the Department of the Environment, Water, Heritage and the Arts) and its predecessors. A number of projects have been supported from other sources. For example, the major blue whale study in the Perth Canyon was funded by the Australian Defence Department, while the Queensland University HARC humpback study has been funded jointly by the US Office of Naval Research and the

(Australian) Defence Science and Technology Organisation (DSTO). Smaller projects have been funded by bodies such as the Australian Marine Mammal Research Centre and the Winifred Scott Foundation.

The major support for Australian marine mammal research, particularly for longer term studies, has recently come from the Natural Heritage Trust through the federal government department responsible for the environment. The situation has now changed somewhat with the establishment in 2006 of the Australian Centre for Applied Marine Mammal Science (ACAMMS). Based in Hobart, under the leadership of Dr Nick Gales, ACAMMS focuses on understanding, protecting and conserving whales, dolphins and other marine mammals of the Australian region. In particular ACAMMS has responsibility, via an Advisory Committee, for recommending to the current Australian Government Department of the Environment, Water, Heritage and the Arts, projects for funding by the Commonwealth Environment Research Facility (CERF). At its inception, ACCAMS had a number of priorities for projects to be considered for funding, within the general framework of populations already listed as threatened and where human activities are most likely to have an effect at the population level. The priorities were, in order: a) quantifying population status and dynamics; b) describing threats, particularly anthropogenic ones; c) developing methods for risk management and mitigation; and d) developing new technologies. For 2007–2008, there was emphasis on projects that would be most likely to lead to major conservation benefits for marine mammals and, indirectly, other threatened marine species, and that include collaboration across disciplines. Of 11 projects funded, five involved cetaceans, all on great whales. Of these, four were on humpbacks – two population studies, one to develop a computerised fluke photograph matching system, and one on the impact of noise. The fifth was a population study of pygmy blue whales. Hopefully, support for a mix of 'cutting edge', shorter term projects and longer term studies of direct relevance to the requirements of the recovery plans for the relevant species, will continue and even expand.

Currently under the EPBC Act, blue whales and southern right whales are considered the most threatened of Australia's great whales. Both are listed as Endangered. In the next category, those considered Vulnerable are sei, fin and humpbacks. Even though both seem to be increasing, at or close to their biological maximum, southern rights and humpbacks are presumably listed differently because the former are still at only a small fraction of their original numbers, that is, before whaling, whereas the latter are currently well on the way to that level, and may even reach it within only a few years. Even with the special permit catches of minke, fin and humpback – all at present, or as proposed, at levels unlikely to have any major effect, if at all, on those species' populations[5] – there seems no likely foreseeable threat to any Australian great whale population from commercial whaling. If commercial whaling were to resume, it should be within the strict confines of a management scheme embodying the principles of the IWC's Revised Management Procedure, and thus bolstered against any possibility of excessive catches. Other factors, including development, climate change and possibly even harvesting of prey, are currently more threatening. A thoughtful article by Dr Randy Reeves and Dr Peter Reijnders (see quote at the start of this chapter), written in 2002 but equally valid today, sets such current conservation practices as lists of threatened species, recovery plans, protected areas and so forth – all supported by governments and NGOs – against what they term 'today's accelerating drive to deregulate and promote free enterprise'. In that climate, Reeves and Reijnders reflect the concerns today being aired freely, for instance in the context of growing public awareness of climate change, that human population increase and growth in per capita consumption must be checked, even reversed, if wild populations, including those of marine mammals, are not to be doomed. They pose the question, 'Why should the survival of a wild species take precedence over the satisfaction of a particular array of human needs and desires?', and believe the importance of conservation cannot be supported by science alone, but needs an ethical and moral dimension.

Where then does that leave Australia's great whales? Certainly there is at present a major groundswell of public opinion against anything even remotely seen as harming any one individual or population. The hope is that it will be possible, within that context and the current legislative framework, to ensure that the great whales continue, in the words of the 1996 Australian Cetacean Action Plan, to 'remain secure from preventable pressures, able to live their lives accordingly while remaining a continuing source of interest, wonder and enjoyment for the future'.

---

5   Except perhaps for humpback removals in the eastern part of Antarctic Area V (that is, between ca 150°E and 170°W) which could adversely affect the currently very small remnant breeding populations in the central South Pacific (see earlier in this chapter).

GLOSSARY

ACAMMS   the Australian Centre for Applied Marine Mammal Science; established in 2006, for understanding, protecting and conserving whales, dolphins and other marine mammals of the Australian region; based in Hobart, Tasmania (www.aad.gov.au).

ambergris   a waxy concretion formed in the sperm whale intestine, once important in perfume manufacture.

Antarctic Circumpolar Current   the most powerful current in the world, circling the globe in the **Southern Ocean** between 40°S and 60°S; formerly known as the West Wind Drift.

Antarctic Convergence   an oceanic zone approximately 20–30 nautical miles (35–55 km) wide separating cold Antarctic water from warmer subantarctic water, with a change in sea surface temperature (**SST**) from ca 3°–6°C (also called **Polar Front**).

anthropogenic   arising from human activity, e.g. noise.

archaeocetes (archaeoceti)   one of the three cetacean suborders; now extinct.

artiodactyls   the order of hoofed mammals including pigs, cattle, hippopotamuses.

bachelor group   an aggregation of maturing male sperm whales, usually of similar age, travelling together.

balaenopterid   a member of the family Balaenopteridae (see **rorqual**).

baleen   keratinous plates hanging in rows from the upper jaw of mysticete (baleen, or whalebone) whales, forming the filter-feeding apparatus; once commercially important for strength and flexibility, as in corset stays, umbrella spokes (see **whalebone**).

baleen whales   a member of the suborder Mysticeti – whalebone whales.

bay whaling   traditional (**open boat**) whaling conducted coastally from vessels in selected bays (see **shore whaling**), typically (in Australian waters) on southern right and humpback whales.

beaching   the act of a whale **stranding** on the coastline, but including carcasses washed ashore.

beaked whales cetaceans of the family Ziphiidae, characterised by a prominent **rostrum** and highly reduced dentition (usually one or two teeth, only in the lower jaw, only erupting in the male); most are deep divers, many are very rare, some being known from only a few specimens.

beluga the white whale, *Delphinapterus leucas*; resident in Arctic and sub-Arctic waters, of the family Monodontidae.

the bends decompression sickness; where inert gases, usually nitrogen, dissolved in blood or other body fluids and tissues, come out of solution as bubbles, for example when a diver ascends too rapidly from depth.

bioluminescence light emission by some marine organisms, e.g. squid, derived from a chemical reaction; used to confuse predators, or to attract prey or mates.

blanket piece a spiral portion of blubber stripped from a whale carcass alongside a **whaleship** during open boat (traditional) whaling, hauled inboard and cut up for **trying out** (see **horse pieces, try pots**).

blow also 'spout', the vapour cloud formed when a whale exhales ('blows') at the surface; its shape is distinctive for various species – tall and columnar (blue whales, fin whales), shorter and bushy (humpbacks), V-shaped (right whales), forward, low and diagonal (sperm whales).

blowhole the whale's nostril/s – single in **toothed whales**, double in **baleen whales**.

blubber the fatty layer immediately underlying the skin in cetaceans and other marine mammals; important for insulation and as a food store.

breaching leaping partly or wholly out of the water, usually falling sideways or back, creating a large and easily visible splash; common for example in humpbacks.

bubble feeding, bubble net a characteristic of humpback whales where the animal emits a stream of bubbles, often while swimming in a circle, to entrap prey.

BWU the blue whale unit; first devised in the 1930s, later adopted by the **IWC** for allocation of catch limits, based on oil yields: one bwu = one blue whale, two fin whales, 2.5 humpbacks, six sei.

callosities patches of thickened, wart-like skin on the heads of right whales, usually infested by **cyamids**; their individually distinct patterns form the basis for photographic identification of right whales.

callus   a thickened patch of skin at the front of the sperm whale **dorsal hump** present in adult females.

capelin   small **pelagic** shoaling fish, common in the northern hemisphere and food for humpback whales, e.g. off eastern Canada and north-east USA.

case   the upper part of the sperm whale head, a vast reservoir containing **spermaceti**, lying above the **junk**.

caudal peduncle   the part of the cetacean body behind the **dorsal fin**, bearing the tail **flukes**; often flattened laterally, in fin whales so much so as to give them the (now largely disused) name '**razorback**'.

CCAMLR   the Convention on the Conservation of Antarctic Marine Living Resources, under which the Commission on Conservation of Antarctic Marine Living Resources (also CCAMLR), established in 1982, manages living marine resources in the Antarctic, based on an ecosystem approach. Its headquarters are in Hobart, Tasmania (www.ccamlr.org).

CERF   the (Australian) Commonwealth Environment Research Facility, from which **ACAMMS** has obtained substantial funding to support marine mammal research.

cetaceans   animals in the order Cetacea; the whales, dolphins and porpoises.

CITES regulations   regulations under the Convention on International Trade in Endangered Species of Flora and Fauna, 1975; designed to ensure that international trade in specimens of wild animals and plants does not threaten their survival. Import and export of cetacean products are subject to its provisions.

CLA   the Catch Limit Algorithm, for calculating catch limits for commercial whaling under the **IWC**'s Revised Management Procedure (**RMP**) within the (yet to be adopted) Revised Management Scheme (**RMS**).

clicks, codas, creaks   noises associated with toothed whales, particularly sperm whales.

climate change   global warming; used nowadays in the context of human induced temperature increase arising from increased levels of carbon dioxide in the atmosphere.

coalfish/saithe   a common northern European fish, related to the pollack, often associated with sei whales off Norway, and from which sei whales get their common name.

Coast of New Holland Ground a favoured area for traditional sperm and right whaling off Western Australia (see **whaling grounds**).

continental shelf/slope the submerged border of a continent, approximately 200 metres deep, bordered by the continental slope which itself leads, often abruptly, to the ocean bottom.

cookie-cutter sharks small warm water sharks (*Isistius*), the origin of whose round 'ice cream scoop' bites on whale bodies were long a mystery to biologists; their healing scars can create a 'galvanised iron' sheen to the skin of heavily scarred animals, e.g. sei whales.

copepods small planktonic crustaceans, the food of some whales, particularly right and sei whales.

cyamids whale lice; small parasitic/commensal crustaceans frequently found on whale skin, especially in right whale **callosities**.

Department of the Environment, Water, Heritage and the Arts Australian federal department (formerly the Department of the Environment and Water Resources) responsible for the EPBC Act (www.environment.gov.au).

diatom film a yellowish bloom of marine diatoms (*Cocconeis*) that forms on the skin of whales in the Antarctic; its extent can be used as a rough gauge of the length of time an animal has been in colder waters. Its presence on the underside of blue whales gave rise to the now archaic name '**sulphur bottom**' for that species.

dolphins small cetaceans less than about four metres long, distinguished from **porpoises** by their (usually) distinct beak, size (more than ca 2 metres long) and pointed teeth.

dorsal fin/hump the fin on the middle of the back of cetaceans, absent in some (e.g. right whales) and modified as a hump in sperm whales.

earhole in whales, closed and inconspicuous at the surface; there is no pinna (external ear) (see **external auditory meatus**).

earplug a waxy body occupying the **external auditory meatus** of **baleen whales**, formed in layers, used in age determination (see **growth layer group**).

echolocation the use of high-frequency sound to locate objects and investigate the surroundings, as employed by toothed whales.

El Niño originally a weak warm current occurring around Christmas time (El Niño = male child) along the coasts of Ecuador and Peru, leading to dramatic changes in water conditions and fish and plant life; particularly strong events, some persisting more than one year, have occurred recently, with worldwide atmospheric consequences.

EPBC Act   the *Environment Protection and Biodiversity Conservation Act 1999*; administered by the Department of the Environment and Water Resources (now the Department of the Environment, Water, Heritage and the Arts).

euphausiids   see **krill**.

external auditory meatus   the canal from the **earhole** to the eardrum; the canal is long and (in baleen whales) occupied by the **earplug**.

fasten   the act of 'fastening' to a whale by means of a **harpoon**.

fid   a tapered tool used in splicing ropes, often made from sperm whale skeletal bone.

filter feeding   as employed by **baleen whales**, straining food from sea water through the **baleen**.

flipper   the forelimb of, for example, a whale, dolphin, seal; preferable to 'pec' (short for pectoral), or 'pec fin', recently in common use in some circles.

fluke/s   the flattened, horizontal, cetacean tail; the black and white pattern on the underside of humpback flukes is used in photographic identification of individual animals.

fluke up/fluking   lifting the tail out of the water on diving; common in humpbacks and sperm whales.

flurry   whaler's term for the whale's death throes while killing with the **lance**.

Frost Inquiry   the Judicial Inquiry under Sir Sidney Frost established by the Australian Government in 1978 to enquire into the Australian whaling industry.

the Georges Report   the report of the 1985 Australian Senate Select Review inquiring into the keeping of marine mammals, particularly dolphins, in captivity.

great whales   the six largest baleen whales (blue, fin, sei, Bryde's, humpback, southern right) and the sperm whale; for the purposes of this book the minke whale has also been included.

Greenland whale   a former name for the eastern Arctic bowhead whale, found in Greenland seas.

growth layer group (GLG) a repeating pattern of growth layers used to determine an animal's age, found in tissues such as baleen whale **earplugs** and sperm whale teeth.

harpoon a spear-like implement comprising a barbed head and an iron shaft fitted with a longer wooden shaft, used to **'fasten'** a whale during capture.

horse pieces small chunks of **blubber** cut from the **blanket piece** prior to **trying out** on a **whaleship**.

IMS Initial Management Stock, as classified by the **IWC** under its New Management Procedure (**NMP**).

International Observer Scheme appointment of independent observers by the **IWC** to police whaling operations; introduced in 1972.

iron a whaling **harpoon**.

IUCN the World Conservation Union, formerly the International Union for the Conservation of Nature; responsible for the 'Red List' assessments of the status of threatened species.

IWC the International Whaling Commission, established under the International Convention for the Regulation of Whaling, 1946; responsible for 'the proper conservation of whale stocks' and 'the orderly development of the whaling industry'.

JARPA the Japanese Research Programme in the Antarctic, under which some 400 Antarctic minke whales were killed under a special scientific catch permit (issued under the provisions of the International Whaling Convention which established the International Whaling Commission – **IWC**) each year from the summer season 1987–1988, alternately in Antarctic Areas IV and V, i.e. south of Australia between 70°E–130°E and 130°E–170°W respectively.

JARPA II the sequel to **JARPA**, begun in 2005–2006, under which the minke catch was increased to a maximum per year of 850, plus or minus an allowance of 10 per cent, and 10 fin whales, for a two-year feasibility study. From 2007–2008, catches were to be extended to include 50 humpbacks and 50 fin whales, although under public pressure the plan to catch humpbacks was not implemented.

junk the lower part of the sperm whale head, honeycombed with fibrous tissue, situated below the **case**.

***K* strategists** animals that tend to grow slowly, have few young, a long life span and exhibit extended parental care, in contrast to *r* **strategists**.

killing iron see **lance**

krill small shrimp-like marine crustaceans; Antarctic krill (*Euphausia superba*) are the staple diet of **baleen whales** feeding in the Antarctic. Another species (*E. crystallorophias*) is food for some species (e.g. blue, minke) near the ice, and warmer water species (*E. recurva, Nyctiphanes australis*) are food for pygmy blue whales, e.g. in the Perth Canyon, Western Australia, and in the Bonney Upwelling, Victoria.

lance a sharp bladed iron spear with wooden shaft, for plunging into a harpooned animal to kill it (see **iron, flurry**).

larynx the voice box; a cartilaginous and muscular enlargement of the windpipe (trachea), containing the **vocal cords**.

leviathan a large aquatic monster, synonymous with the whale.

lobtailing behaviour where a whale slaps its flukes repeatedly on the water.

logging describes a whale (typically a right or sperm whale) lying quietly at the surface.

lunge feeding as in most **rorquals**, where they lunge into a school of prey, gulping in great quantities of water and straining the food out through the fringing 'doormat' of the baleen plates, in contrast to **skimming**.

median dorsal ridge a raised line running the length of the **rostrum**, diagnostic (singly) of fin and sei whales, and (in triplicate) of Bryde's whales.

melon a fatty mass of tissue at the front of the head in **toothed whales**, possibly used to focus sound in **echolocation**.

'Middle Ground' a favoured area for traditional sperm whaling in the Tasman Sea, between eastern Australia and New Zealand (see **whaling grounds**).

Moby-Dick the fictional sperm whale immortalised in Herman Melville's classic novel, first published in London in October 1851 as *The Whale* and in November 1851 in New York as *Moby-Dick; or The Whale* (the hyphen, often not used nowadays, is present in the first (New York) printing).

modern whaling dating from the 1860s with the development of the combination of a fast steam catching boat with an explosive-tipped harpoon mounted on the bow which allowed catching of the fast swimming rorquals (blue, fin and sei whales), previously uncatchable in 'traditional', **open boat whaling**.

moratorium a 'pause' in commercial whaling under which no commercial whaling was to take place; it was introduced by the **IWC** in 1982 to take effect in 1986, and remains in operation.

MSY maximum sustainable yield; the maximum harvest that a resource can sustain in the long term.

myoglobin a blood protein with particular oxygen-binding properties.

mysticetes **whalebone** or **baleen whales** in the suborder Mysticeti.

narwhal *Monodon monoceros*, found only in the Arctic, of the family Monodontidae, characterised by the single long tusk in the male.

Natural Heritage Trust the Australian Government Fund, obtained by the sale of Telstra, from which much government-supported cetacean research has recently been funded through the Department of the Environment, Water, Heritage and the Arts and its predecessors.

New Holland former name for Australia.

'New Holland Ground' a favoured area for traditional sperm and right whaling off the coast of Western Australia (see **whaling grounds**).

NMP adopted by the **IWC** in 1975 for the management of whale stocks, essentially classifying them as Protection (**PS**), Initial (**IMS**) or Sustained (**SMS**) management stocks.

North West Ground a short-lived but prolific right whaling ground off the north-west coast of America (see **whaling grounds**).

nursery group an aggregation of sperm whales consisting of adult females and their young of both sexes.

odontocetes toothed cetaceans in the suborder Odontoceti.

open boat whaling 'traditional' whaling undertaken in historical times, from shore or from (sailing) **whaleships**, hunting the whale from small open **whaleboats**.

pan bone, skeletal bone bone from sperm whales, often used in **scrimshaw** or in shipboard tools such as **fids**.

pelagic in or from the open sea.

Polar Front the **Antarctic Convergence.**

porpoises the smallest cetaceans; usually less than 2 metres long, often without a distinct beak, sometimes lacking a dorsal fin, and with spade-shaped or vestigial teeth.

predator an animal that preys upon other animals for its food.

PS   Protection Stock as classified by the **IWC** under the **NMP**.

***r* strategists**   animals that tend to grow rapidly, have numerous offspring, a short life span, and exhibit little parental care, in contrast with ***K* strategists**.

razorback   an alternative (but little used now) common name for the fin whale, based on the laterally compressed **caudal peduncle**.

recovery plans   prepared under the **EPBC Act** for individual threatened species, to maximise their long-term survival in the wild, by detailing both what has to be done to protect and restore important populations as well as to manage and reduce threats; prepared so far jointly for blue, fin and sei whales, and separately for humpbacks and southern right whales.

RMP   the Revised Management Procedure, adopted by the **IWC**; developed to replace the New Management Procedure **(NMP)** but not yet in force, pending adoption of the Revised Management Scheme (**RMS**).

RMS   the Revised Management Scheme, yet to be adopted by the **IWC**, to administer and control catch limits recommended under its **RMP**.

rorqual   a whale with pleats (**ventral grooves**) in its throat; any of the species in the family Balaenopteridae, including humpbacks.

rostrum   the beak at the front or top of the skull in **cetaceans**; see **beaked whales**.

scrimshaw   originally whalers' handiwork, including tools but also decorated objects, using sperm whale teeth or bone but also baleen, and walrus tusks; usually involving engraved designs highlighted with pigment.

sexual maturity   the average age at which an animal can first reproduce; but see **socially mature**.

shore whaling   whaling conducted from shore in selected bays, typically (in Australian waters) on southern right and humpback whales (see **bay whaling**).

skimming   feeding by, for example, sei whales and right whales, where the whale scoops up a quantity of water as it swims slowly forward through a prey school, in contrast to **lunge feeding**.

SMS   Sustained Management Stock, as classified by the **IWC** under its **NMP**.

socially mature   used of, e.g., an adult male sperm whale which by reason of its social status is able to reproduce; in contrast to sexually mature (see **sexual maturity**).

song as in humpbacks: a series of sounds in a repeated pattern.

Southern Ocean the Antarctic Ocean, circling the globe between the Antarctic continent and ca 60°S; separated from the Indian, Pacific and Atlantic oceans by the **Polar Front (Antarctic Convergence)**.

sperm oil oil obtained from the body of a sperm whale, distinct from **spermaceti**.

spermaceti waxy oil contained in the head of a sperm whale; in contact with air solidifying into a white wax, different from oil from the rest of the body **(sperm oil)** which remains liquid.

spermaceti organ the source of **spermaceti** in the sperm whale head, comprising the **case** and the **junk**.

spermer a **whaleship** voyaging for sperm whales.

spout see **blow**.

spy-hopping raising the head vertically above water, as in humpbacks.

Stockholm Conference the United Nations Conference on the Human Environment, 1972, which set the scene for much of the anti-whaling movement in the later 1970s.

stranding the act of a whale coming ashore, either sick (usually as a single animal), or accidentally (often in a group) (see **beaching**).

Subtropical Front (STF) found at between 35°S and 45°S, marked by a rapid change of surface water temperature from ca 12°C to 7° or 8°C and a distinct decrease in salinity; formerly known as the Subtropical Convergence.

suckling the act of a female mammal feeding its young.

sulphur bottom alternative common name for the blue whale, now in disuse; derived from the yellowish film of diatoms acquired in polar waters (see **diatom film**).

surface active groups (SAGs) groups of adult right whales, often of several males and a single female, interacting vigorously at the surface.

swim bladder gas-filled internal organ of a fish used in controlling buoyancy.

telomeres the ends of a chromosome that may shorten or degrade during life and have been thought of as possible indicators of an animal's age.

toothed whales the **odontocetes**, including sperm whales, dolphins and porpoises.

try pots   large iron pots or 'kettles' set in a brick base (as the **try works**) on shore or on a **whaleship**, in which **blubber** was rendered down for oil.

try works   a combination of **try pots** on a brick base, used in rendering **blubber** for oil, often fuelled by scraps of blubber from which the oil has already been extracted.

trying out   the process of treating **blubber** in **try pots** to extract oil.

tympanic bullae   the earbones; growth layers in the bone have been used for whale age determination but with limited success.

'unaccompanied' animals   adult (or subadult) right whales of either sex 'unaccompanied' by calves.

ungulates   hoofed mammals of the order Perrissodactyla; 'odd-toed', including horses, tapirs and rhinoceroses.

upwelling   wind-driven movement of colder water from deeper regions; important for bringing nutrients to the surface.

Van Diemen's Land   original name for Tasmania; named in 1642 by Dutch explorer Abel Tasman after the then Governor-General of the Dutch East Indies.

ventral grooves (ventral pleats)   parallel grooves or pleats running from the chin towards the navel in **rorquals**, allowing great distension of the mouth in feeding.

vestibular sac   a broad flat sac in **odontocetes** (including sperm whales), just below the **blowhole**; in sound production air can be recycled through the sac or be released as bubbles into the water.

vocal cords   small muscular bands in the **larynx** that in humans vibrate to produce the voice.

whale lice   cyamid crustaceans parasitic on whale skin, especially in right whale head **callosities**.

whale oil   the oil obtained from baleen whales, as opposed to **sperm oil**.

whaleboat   a small (ca nine metre) clinker-built, light, double-ended craft, rowed or sailed up to a whale in traditional (**open boat**) **whaling**.

whalebone   colloquial name for **baleen**.

whalecraft   traditional implements used to **fasten** and kill whales, e.g. **harpoons, lances**, as carried in a **whaleboat**.

whaleship   a sailing vessel equipped for whaling in traditional **open boat whaling**.

whaling grounds   favoured areas of ocean where whalers, particularly in traditional open boat whaling, expected to find concentrations of whales, especially sperm, right and humpback whales.   .

Yankee whalers   American whalers particularly from New England, USA, who pioneered and dominated much of traditional (**open boat**) **whaling** from the early eighteenth century.

FURTHER READING

Much of the extensive literature on whales and whaling is in scientific journals, some not readily available. A number are now on the internet but many are accessible to subscribers only. The two major specialist journals are *Marine Mammal Science* and the *Journal of Cetacean Research and Management.*

For up-to-date listing of genera and species I have relied on Jim Mead and Bob Brownell's 2005 *Order Cetacea*. Though published in 1983, I still recommend the *Sierra Club Handbook* as the best world guide; the more recent (2002) National Audubon Society's *Guide to Marine Mammals of the World* (with the same senior author as the Sierra Club handbook) is very comprehensive. Dr Neil Mackintosh's *The Stocks of Whales* remains, in my opinion, the most readable and comprehensive book on whales and the whaling industry even though only to that date (1965), and should be required reading for all who aspire to an understanding of the subject. L Harrison Matthews' *The Whale* has much of interest, particularly on mythology and folklore, as well as on products. Annalisa Berta and James L Sumich's *Marine Mammals: Evolutionary Biology* is technical but illuminating, and I've found much to ponder in *Marine Mammal Biology, An Evolutionary Approach*, edited by Russ Hoelzel, especially the final article – *Conservation and Management* – by Randy Reeves and Peter Reijnders. *Dugongs, Whales, Dolphins and Seals* by Michael Bryden, Helene Marsh and Peter Shaughnessy covers Australia's marine mammal fauna as a whole, as does the just released third edition of the *Mammals of Australia*, published by the Queensland Museum. Alan Baker's *Identification Guide* covers cetaceans of the Australasian region. Professor WJ Dakin's *Whalemen Adventurers* remains the classic history – Graham Chittleborough's introduction to the 1963 Sirius Books edition brings it forward from the 1930s to the 1960s. The 1978 Frost Report – the turning point in Australia's attitude to whaling – provides a most lucid and comprehensive account of the Australian situation to that date. The 1996 *Action Plan for Australian Cetaceans* with its individual species synopses still contains much useful information. Graham Chittleborough's classic work on the two Australian humpback populations, and Bill Dawbin's two papers, one on early right whaling in the south-west Pacific (including south-east Australia) and the other on humpback migration, show how much we owe to them both. My little book on Western Australia's humpback and right whales gives the background to and details of the recovery of those two species off the west coast to the mid 1990s, and has formed the basis for much of my approach here. The Australian Petroleum Production and Exploration Association (APPEA)'s CD *Search Australian Whales and Dolphins* combines a comprehensive field guide and aid to reporting sightings. Pete Gill and Cecilia Burke's guide to *Whale Watching in Australian and New Zealand Waters* highlights the best places to see whales in our waters.

Current Australian legislation, and many other relevant matters, for example, whalewatching, are fully covered on the website of the Department of the Environment, Water, Heritage and the Arts (www.environment.gov.au).

Peter Best's recently published *Whales and Dolphins of the Southern African Sub-region* comprehensively describes the extensive cetacean fauna of our neighbouring continent across the Indian Ocean.

But for marine mammals as a whole, and in the current context particularly for the larger cetaceans, one cannot go beyond the *Encyclopedia of Marine Mammals*. It has been by my side throughout the writing of this book. Particular articles I have relied on, in addition to the various species accounts, include those on aquaria/oceanaria, ambergris, baleen whales, early and modern whaling, intelligence, whalewatching, the International Whaling Commission, scrimshaw, sound production and strandings. I hope the individual authors will feel I have done them justice.

Recommended titles are listed below and include the sources of the various figures reproduced in the text.

Baker AN (1983). *Whales and Dolphins of New Zealand and Australia: An Identification Guide.* Victoria University Press: Wellington.
Bannister JL (1968). An aerial survey for sperm whales off the coast of Western Australia 1963–65. *Australian Journal of Marine and Freshwater Research* **19** (1): 31–51.
Bannister JL (1995). *Western Australian Humpback and Right Whales: An Increasing Success Story.* Western Australian Museum: Perth.
Bannister JL (2001). Status of southern right whales (*Eubalaena australis*) off Australia. *Journal of Cetacean Research and Management* **(Special Issue 2)**: 103–10.
Bannister JL (2008). Population dynamics of right whales off southern Australia, 2007. Final report to the Department for the Environment, Water, Heritage and the Arts, Canberra (available from the Department – GPO Box 797, Canberra ACT 2600, Australia).

Bannister JL and Hedley SL (2001). Southern Hemisphere Group IV humpback whales: Their status from recent aerial survey. *Memoirs of the Queensland Museum* **47** (2): 587–98.

Bannister JL, Kemper CM and Warneke RM (1996). *The Action Plan for Australian Cetaceans*. Wildlife Australia, Endangered Species Program, Project Number 380. Australian Nature Conservation Agency: Canberra.

Beale T (1839). *The Natural History of the Sperm Whale*. John Van Voorst: London.

Berta A and Sumich JL (1999). *Marine Mammals: Evolutionary Biology*. Academic Press: San Diego.

Best PB (2007). *Whales and Dolphins of the Southern African Sub-region*. Cambridge University Press: Cape Town.

Best PB, Bannister JL, Brownell RL Jr and Donovan GP (eds) (2001). Right whales: worldwide status. *Journal of Cetacean Research and Management* **(Special Issue 2)**.

Bryden M, Marsh M and Shaughnessy P (1998). *Dugongs, Whales, Dolphins and Seals: A Guide to the Sea Mammals of Australasia*. Allen and Unwin: Sydney.

Burnell SR (2001). Aspects of the reproductive biology, movements and site fidelity of right whales off Australia. *Journal of Cetacean Research and Management* **(Special Issue 2)**: 89–102.

Chittleborough RG (1965). Dynamics of two populations of the humpback whale, *Megaptera novaeangliae* (Borowski). *Australian Journal of Marine and Freshwater Research* **16**: 33–128.

Dakin WJ (1963). *Whalemen Adventurers. The Story of Whaling in Australian Waters and other Southern Seas Related Thereto, from the Days of Sails to Modern Times*. Sirius Books (Revised edition). Angus and Robertson: Sydney.

Dawbin WH (1966). The seasonal migratory cycle of humpback whales. In *Whales, Dolphins and Porpoises*. (ed. KR Norris) pp. 145–70. University of California Press: Berkeley and Los Angeles.

Dawbin WH (1986). Right whales caught in waters around south-eastern Australia and New Zealand during the nineteenth and early twentieth centuries. In *Right Whales: Past and Present Status*. (eds RL Brownell, PB Best and JH Prescott) pp. 261–68. Reports of the International Whaling Commission, Special Issue 10.

Frost S (1978). *Whales and Whaling. Report of the Independent Inquiry conducted by Sir Sydney Frost*. Vol 1, The Report; Vol 2, Commissioned Papers. Australian Government Publishing Service: Canberra.

Gill PC and Burke C (2004). *Whale Watching in Australia and New Zealand Waters*. Reed New Holland: Sydney.

Hoelzel AR (ed.) (2002). *Marine Mammal Biology: An Evolutionary Approach*. Blackwell Science: Oxford.

Leatherwood S and Reeves RR (1983). *The Sierra Club Handbook of Whales and Dolphins*. Sierra Club Books: San Francisco.

Mackintosh NA (1965). *The Stocks of Whales*. Fishing News (Books) Ltd: London.

Matthews LH (1968). *The Whale*. George Allen and Unwin: London.

Mead JG and Brownell RLJ (2005). Order Cetacea. In *Mammal Species of the World*. (Third edition) 2 Volumes. (eds DE Wilson and DM Reader) pp. 723–45. Johns Hopkins University Press, Baltimore.

Melville H (1851). *Moby-Dick or the Whale*. Harper and Brothers: New York.

Noad M, Paton D and Cato DH (2006). Absolute and relative abundance of Australian east coast humpback whales *Megaptera novaeangliae*. Document SC/A06/HW27 presented to the Comprehensive Assessment of Southern Hemisphere Humpback Whales, Hobart, Tasmania, April 2006, (available from the International Whaling Commission, The Red House, 135 Station Road, Impington, Cambridge CB24 9NP, UK).

Paxton CGM, Hedley SL and Bannister JL (2006). Group IV humpback whales: their status from aerial and land-based surveys off Western Australia, 2005. Document SC/A06/HW3 presented to the Comprehensive Assessment of Southern Hemisphere Humpback Whales, Hobart, Tasmania, April 2006, (available from the International Whaling Commission, The Red House, 135 Station Road, Impington, Cambridge CB24 9NP, UK).

Perrin WF, Wursig B and Thewissen JGM (eds) (2002). *Encyclopedia of Marine Mammals*. Academic Press: San Diego.

Reeves RR and Reijnders PJH (2002). Conservation and management. In *Marine Mammal Biology: An Evolutionary Approach*. (ed. AR Hoelzel) pp. 388–415. Blackwell Science: Oxford.

Reeves RR, Stewart BS, Clapham PJ and Powell J (2002). *National Audubon Guide to Marine Mammals*. Alfred A Knopf: New York.

Rice DW (1998). *Marine Mammals of the World, Systematics and Distribution*. The Society for Marine Mammalogy, Special Publication Number 4.

Van Dyck S and Strahan R (eds) (2008). *The Mammals of Australia*. (Third edition). Reed New Holland: Sydney.

INDEX

Figure references are in **bold**

ACAMMS 121
Action Plan for Australian Cetaceans 120, 122
ambergris 6, 11–12, 22, 23, **63**
Antarctic blue whale *see* blue whale
Antarctic circumpolar current 34
Antarctic Convergence *see* Polar Front
Antarctic minke whale *see* minke whale
archaeocetes 28
Archaeoceti *see* archaeocetes
artiodactyls 28
Australian Centre for Applied Marine Mammal Science *see* ACAMMS
Australian cetacean fauna 2–3

bachelor group 101
Balaenidae 3, 4 *see also* right whales
Balaenoptera 4 *see also* baleen whale
    acutorostrata see common minke whale
    bonaerensis see Antarctic minke whake
    borealis see sei whale
    brydei see Bryde's whale
    edeni see Bryde's whale
    musculus see Antarctic blue whale, pygmy blue whale
    m. brevicauda see pygmy blue whale
    m. intermedia see Antarctic blue whale
    omurai see Omura's whale
    physalus see fin whale
balaenopterid *see* Balaenopteridae
Balaenopteridae 3, 4
baleen 1, 2, 12, 14, 18, 22, 28, 29, **32**, 38, 43, 46, 50, 53, 56, 57, 59, 71, 73, 75, 77, 82, 89, 93, 94 *see also* whalebone
baleen plates *see* baleen
baleen whale 1, 3, 4, 7, 8, 22, 32, 33, 34, 36, 37, 41–3, 44, 46 *see also* Balaenopteridae
bay whaling **19**
beaked whales 2, 7, 9, 28
beluga 2, 24, 28, 29
the bends 31
blanket piece **18**
blow/blowing 7, 38, 50, 51, 56, 59, **64**, 74, 89, **90**, 91, 98, 99, 102
blowhole 6, 7, 28, **30**, 37, 98, **99**, 100
blubber 14, **18**, **19**, 20, 21, 22, 31, 45
blue whale 4, 6, 7, 20, 21, 24, 29, **30**, 33, 34, **35**, 36, 38, 39, 41, 43, 44, **50**, 50–5, 56, 57, 58, 60, **64**, **65**, 74, 107, 117, 121

Antarctic blue 3, 21, 39, 40, 41, 50, 51, 52, 53, 54, 55, **64**
breeding 54
distribution and movements 51–3
feeding and food 53–4
pygmy blue 3, 4, 35, 38, 39, 40, 41, 42, 43, 50, 51, 52, 53, 54, **64**, 75, 110, 121
status 54–5
Blue Whale Unit *see* bwu
Bonney Upwelling 35, 41, 53, 54, 55, 60
bottlenose whale *see Hyperoodon*
breaching 78, 90
breeding grounds 34, 39, 41, 45, 54, 57, 58, 60, 74, 78, 79, 82, 88, 92 *see also* calving grounds
Breeding Stocks, of humpbacks 79–80, **81**, 84, 85, 87, 117
Bryde's whale 29, 34, 35, 38, 42, 43, 59, 60, 61, **62**, **66**, 71–2, 119
breeding 72
distribution and movements 71
feeding and food 71–2
status 72
bubble feeding 81
bubble net 81
bwu 107

callosities 45, **68**, 89, **91**
callus 99
calving, calving rate 46, 54, 58, 60, 61, 72, 75, 81, 82, 93, 94, 103, 118, 119
calving grounds 34, 45, 79 *see also* breeding grounds
capelin 57
Caperea see pygmy right whale
case 18, 22, 28, 99
Catch Limit Algorithm *see* CLA
catch limits 107–108 *see also* quotas
CCAMLR 119, 120
Cephalorhynchus 2
CERF 121
Cetacea 1, 3, 28
Cetaceans 1, 2, 7, 8, 9, 23, 24, 25, 28, 30, 33, 37, 109, 110, 116, 117, 119, 121
Chilean dolphin *see Cephalorhynchus*
CITES 12
CLA 111, 112
climate change 44, 118, 119, 120, 122
coalfish *see* saithe
Coast of New Holland Ground 15, 103
cohorts, in right whales 97

Commerson's dolphin *see Cephalorhynchus*
Commission for the Conservation of Antarctic Living Resources *see* CCAMLR
'Committee of Three' *see* Special Committee of Three Scientists
Commonwealth Environment Research Facility *see* CERF
competition 44, 60 *see also* sperm competition
continental shelf 42, 44, 53, 100, **101**
continental slope 42, 102
Convention on the Conservation of Antarctic Living Resources *see* CCAMLR
Convention on International Trade in Endangered Species *see* CITES
cookie-cutter shark 59 *see also Isistius*
copepods 43, 60, 71, 92, 93
cultural evolution 38, 80
cyamids 45 *see also* whale lice

Department of the Environment, Water, Heritage and the Arts *see* DEWHA
DEWHA 25, 120, 121
diatom film 45, 51
dolphins 1, 2, 3, 5, 6, 7, 8, 24, 28, 36, 37, 39, 44, 75, 100, 121
dorsal fin **30**, 50, 56, 59, **65**, **66**, **68**, 70, 73, 77, 89, 91, 98
dorsal hump **30**, **70**, 98, 99
dwarf minke whale *see* minke whale

earplug 46, 82
echolocation 28, 38, 44, 100
echosonogram **35**
El Niño 94, 97
Environment Protection and Biodiversity Conservation Act *see* EPBC Act
EPBC Act 49, 55, 58, 61, 88, 97, 110, 116, 117, 121
Eubalaena australis see southern right whale
Euphausia see also euphausiids
　crystallorophias 75
　recurva 54
　superba 54, 57, 75, 80, 93
euphausiids 43, 54, 71, 75 *see also Euphausia,* krill, *Nyctiphanes*
evolution 28, 38, 80 *see also* cultural evolution
evolution and special adaptations 28–32
evolutionary tree **29**
external auditory meatus 46

fastened 17
feeding grounds 34, 36, 41, 46, 57, 59, 60, 61, 74, 76, 78, 80, 81, 82, 92, 93, 94, 117, 119
fid **63**

filter feeding 1, 42, 43, 60, 94
fin whale 3, 6, 8, 20, **32**, 33, 34, 35, 36, 37, 38, 39, 41, 44, 46, 53, **56**, 56–8, 59, 60, **65**, 66, 82, 107, 113, 118
　breeding 58
　distribution and movements 57
　feeding and food 57
　status 58
flipper **30**, **67**, 73, 90
fluke **30**, 31, 37, 38, 45, 50, 59, 77, 82, **83**, 89, 90, 98, 102, 121
fluke up 56, 78, 98, 102
fluking *see* fluke up
flurry 18
food and feeding 42–4
freshwater river dolphins 2
Frost Inquiry 106, 109, 111

Georges Report 24
gestation 58, 60, 72, 75, 81, 94, 103
GLG 46, 47, 82
gray whales 2, 14, 24, 29, 107
grazers 7
great whales 3, 4, 24, 28, 29, 33, 37, 41, 46, 52, 102, 117, 119, 120, 121, 122
Greenland whale 14
Ground *see* whaling ground
growth and reproduction 46–7
growth layer group *see* GLG
gulp feeding, gulping 43, 53, 57, 75, 81

habitat 41–2
harpoon 17, 20, **21** *see also* iron
Haviside's dolphin *see Cephalorhynchus*
Hector's dolphin *see Cephalorhynchus*
horse pieces 18
humpback whale 2, 3, 4, 7, 10, 14, 19, 20, 24, **25**, 29, 33, 34, 35, 37, 38, 39, 40, 41, 43, 44, 45, 46, 52, 53, 57, 58, 60, 61, **68**, 75, 77–88, **79**, **83**, **87**, 95, 103, 106, 107, 108, 110, 112, 113, 116, 117, 118, 119, 120, 121, 122
　breeding 81–2
　distribution and movements 78–80
　feeding and food 80–1
　status 83–8
Hyperoodon 2

IMS 111
Indohyus 28
Indopacetus 2
Initial Management Stock *see* IMS
intelligence 8
International Convention for the Regulation of Whaling *see* IWC
International Observer Scheme 107, 108, 111

International Union for the Conservation of Nature *see* IUCN, red list
International Whaling Commission 25, 72, 79, 84, 85, 103, 107, 108 *see also* IWC
iron 17 *see also* harpoon, killing iron, lance
Isistius 45, 74 *see also* cookie-cutter shark
IUCN 49, 55, 58, 61, 88, 97, 104
    red list 49
IWC 107, 108, 109, 110, 111, 120, 122 *see also* International Whaling Commission

Japanese Research Program in the Antarctic *see* JARPA, JARPA II
JARPA 76, 113
JARPA II 44, 113, 117
junk 18, 22

K strategists 33
killing iron 18
krill 41, 42, 43, 53, 54, 57, 60, 65, 75, 80, 93, 119 *see also Euphausia*, euphausiids, *Nyctiphanes*

lactation, lactating females 43, 46, 57
lance 18 *see also* killing iron
larynx 32, 37
leviathan 5, 6
life history 33–47
lobtailing 90
logging 45, 92
lunge feeding, lunging 43, 53, 57, 60, 75, 80, 81

mating 46, 54, 58, 60, 72, 79, 81, 94
matriarchies 8
maximum sustainable yield *see* MSY
median dorsal ridge 56
Megaptera novaeangliae see humpback whale
melon 28, 100
Middle Ground 103
migration 34-6
minke whale 73–76
    Antarctic minke 3, 4, 29, **67**, **73**, 73, 74, 75, 76, 113, 119, 120
    breeding 75–6
    common minke 67, 73, 74
    distribution and movements 74–5
    dwarf minke 1, 3, 4, 25, 38, 41, **67**, **73**, 74, 75, 119
    feeding and food 75
    status 76
Moby-Dick 98
modern whaling 21, 61, 83, 103, 106
moratorium 109, 111, 112
MSY 111
myoglobin 32

mysticetes 1, 2, 7, 28 *see also* Mysticeti
Mysticeti 1, 3, 4, 28

narwhal 2, 28
Natural Heritage Trust 120, 121
New Holland 15
New Holland Ground *see* Coast of New Holland Ground
New Management Procedure *see* NMP
nipples 31
NMP 111
'non-consumptive' uses 23–5
North Atlantic right whale 37, 118
North West Ground 15
Nyctiphanes australis 54
nursery group 39, 42, 100, 102

odontocetes 1, 2, 7, 28 *see also* Odontoceti, toothed whales
Odontoceti 1, 3, 28 *see also* odontocetes
Omura's whale 71
open boat whaling 14, 17, **18**, 19, 20, 63, 89, 95, 103, 106
Orcaella heinsohni 2

Perth Canyon 33, 35, 39, 40, 41, 43, 51, 52, 53, 54, 64, 120
Phocoena see spectacled porpoise
phonic lips 37
photoidentification 45
phylogeny *see* evolutionary tree
Physeter macrocephalus see sperm whale
Physeteridae 3 *see also Physeter macrocephalus*
Polar Front 40, 60, 92, 93
porpoises 1, 2, 3, 5, 7, 24, 28
predators and parasites 45
Protection Stock *see* PS
PS 111
pygmy right whale 2, 118

quotas 103, 107–108 *see also* catch limits

r strategists 33
razorback 56
recovery plans 110, 116, 119, 120, 121, 122
research whaling *see* scientific permit whaling, special permit catches
'resting' females 46
Revised Management Procedure *see* RMP
Revised Management Scheme *see* RMS
right whale 22, 25, **30**, 37, 45, 60, 89, 92, 118 *see also* Balaenidae, North Atlantic right whale, southern right whale
RMP 111, 112, 113
RMS 112
rorqual 4, **30**, 46

rostrum 28, 29, 30 *see also* beaked whales

SAGs 37, 94
saithe 59
satellite marking, satellite tagging 41, 53
Scientific Committee (of the International Whaling Commission) 25, 72, 79, 85, 103, 107, 109, 110, 111, 113
scientific permit whaling 3–4, 112, 117, 120 *see also* special permit catches
scrimshaw 11, 12, **63**
sea surface temperature 36, 94, 118, 119
segregation 35–36, 57, 82
sei whale 3, 4, 20, 29, 33, 34, 36, 38, 41, 43, 45, 56, 57, **59**, 59–61, 62, **66**, 71, 72, 76, 107, 110, 111, 116, 119, 120, 121
    breeding 60–1
    distribution and movements 60
    feeding and food 60
    status 61
sexual maturity 33, 54, 58, 61, 76, 82
shore whaling 19, 20 *see also* bay whaling
skimming 43
skimmimg feeding, skimmers 43, 60, 94
SMS 111
snubfin dolphin *see Orcaella heinsohni*
social activity 36–7
socially mature 36, 103
song 40, 78, 80
sound production 37–40
Southern Ocean 11, 44, 118, 119, 120
Southern Ocean Sanctuary 110
southern right whale 2, 3, 4, 29, 33, 34, 44, 45, **68**, **69**, **89**, 89–97, **91**, 110, 116, 118, 119, 120, 121 *see also* right whale, Balaenidae
    breeding 94–5
    distribution and movements 91–3
    feeding and food 93–4
    status 95–7
Special Committee of Three Scientists 107
special permit catches 44, 58, 76, 112, 122 *see also* scientific permit whaling
spectacled porpoise 2
spermaceti 6, 18, 22, 23, 28, 99, 100, 109
spermaceti organ 28, 32, 39, 99, 100
sperm competition 37, 94
sperm oil 15, 22, 23, 109
sperm whale 2, 3, 4, 7, 8, **9**, 10, 11, 12, 14, 15, **17**, **18**, 21, 23, 24, 27, 28, 29, **30**, **31**, 32, 33, 34, 36, 37, 38, 39, 41, 42, 43, 44, 45, 46, 47, **63**, **70**, **98**, 98–104, **99**, 108, 110, 112, 118, 119, 120
    breeding 102–3

distribution and movements 100–2
feeding and food 102
status 103–4
spout *see* blow
spy-hopping 78, **79**
squid, squid beaks 11, 12, 39, 44, 60, 100, 102
Stockholm Conference 111
stranding 8–11, **9**, 52, 72, 99
submarine canyons 42, 102
Subtropical Front 41, 60
suckling 46, 82, 103
'sulphur bottom' 45
surface active groups *see* SAGs
Sustained Management Stock *see* SMS
swimming and diving 33–4

telescoping of the skull 28
telomeres 47
threats 110, 115–120, 121
toothed whales 1, 3, 4, 7, 8, 21, 28, 29, 32, 37, 38, 42, 43, 46, 98 *see also* odontocetes, Odontoceti
tropical bottlenose whale *see Indopacetus*
trying out **19**
try pot 18
tympanic bulla 46

'unaccompanied' animals 93, 94
ungulates 29
upwelling 42

ventral grooves 4, **30**, 37, 43, 53, 56, 57, 59, 62, 71, 75, 77, 81
ventral pleats *see* ventral grooves
vestibular sac 37
vocal cords 37

weaning 54, 58, 60, 72, 76, 81, 94
whaleboat 17
whalebone 6, 7, 11, 20, 22 *see also* baleen
whaleship 16, 17, **18**
whalewatching 4, 5, 24–5, **25**, 78
whale lice 45, 78, 89, 90
whale oil 11, 22
whale products 21–3
whaling 13–21
whaling grounds 15, 17 *see also* Coast of New Holland Ground, Middle Ground, North West Ground

Yankee whalers 4, 14
Yankee whaling 14, 22